ASCE STANDARD

American Society of Civil Engineers

Design Loads on Structures during Construction

This document uses both the International System of Units (SI) and customary units.

Published by the American Society of Civil Engineers

Library of Congress Cataloging-in-Publication Data

Design loads on structures during construction / American Society of Civil Engineers.
pages cm
"ASCE/SEI 37-14."
Includes index.
ISBN 978-0-7844-1309-8 (soft cover : alk. paper) – ISBN 978-0-7844-7869-1 (pdf)
1. Buildings–Standards. 2. Strains and stresses–Standards. I. American Society of Civil Engineers.
TH420.D47 2015
624.1′76–dc23

2014046422

Published by American Society of Civil Engineers
1801 Alexander Bell Drive
Reston, Virginia, 20191-4382
www.asce.org/bookstore | ascelibrary.org

This standard was developed by a consensus standards development process that has been accredited by the American National Standards Institute (ANSI). Accreditation by ANSI, a voluntary accreditation body representing public and private sector standards development organizations in the United States and abroad, signifies that the standards development process used by ASCE has met the ANSI requirements for openness, balance, consensus, and due process.

While ASCE's process is designed to promote standards that reflect a fair and reasoned consensus among all interested participants, while preserving the public health, safety, and welfare that is paramount to its mission, it has not made an independent assessment of and does not warrant the accuracy, completeness, suitability, or utility of any information, apparatus, product, or process discussed herein. ASCE does not intend, nor should anyone interpret, ASCE's standards to replace the sound judgment of a competent professional, having knowledge and experience in the appropriate field(s) of practice, nor to substitute for the standard of care required of such professionals in interpreting and applying the contents of this standard.

ASCE has no authority to enforce compliance with its standards and does not undertake to certify products for compliance or to render any professional services to any person or entity.

ASCE disclaims any and all liability for any personal injury, property damage, financial loss, or other damages of any nature whatsoever, including without limitation any direct, indirect, special, exemplary, or consequential damages, resulting from any person's use of, or reliance on, this standard. Any individual who relies on this standard assumes full responsibility for such use.

Errata: Errata, if any, can be found at http://dx.doi.org/10.1061/9780784413098.

ISBN 978-0-7844-1309-8 (soft cover)
ISBN 978-0-7844-7869-1 (PDF)
Manufactured in the United States of America.

23 22 21 20 19 3 4 5 6 7

ASCE STANDARDS

In 2006, the Board of Direction approved the revision to the ASCE Rules for Standards Committees to govern the writing and maintenance of standards developed by the Society. All such standards are developed by a consensus standards process managed by the Society's Codes and Standards Committee (CSC). The consensus process includes balloting by a balanced standards committee made up of Society members and nonmembers, balloting by the membership of the Society as a whole, and balloting by the public. All standards are updated or reaffirmed by the same process at intervals not exceeding five years.

A complete list of currently available standards is available in the ASCE Library (http://ascelibrary.org/page/books/s-standards).

CONTENTS

PREFACE

The material presented in this publication has been prepared in accordance with recognized engineering principles. This Standard and Commentary should not be used without first securing competent advice with respect to their suitability for any given application. The publication of the material contained herein is not intended as a representation or warranty on the part of the American Society of Civil Engineers, or of any person named herein, that this information is suitable for any general or particular use or promises freedom from infringement of any patent or patents. Anyone making use of this information assumes all liability from such use.

Earlier drafts of this Standard and Commentary were reviewed and balloted several times by the full Standards Committee. The votes and comments returned by the members were reviewed and their proposed resolutions developed by the appropriate subcommittees. The resulting approved changes in the text are included in this volume.

Some of the provisions were adopted from other codes, standards, regulations, and specifications; some reflect prevailing industry design and construction practices; some grew out of the experiences, practices, and opinions of members of the Committee; and some others were developed through research conducted specifically for this Standard by members of the Committee.

Preparation of a standard for *Design Loads on Structures during Construction* and its outline were originally proposed to ASCE by Robert T. Ratay in 1987, resulting in the first edition of the Standard published in 2002 as *ASCE/SEI 37-02, Design Loads on Structures during Construction*. The Committee, through its subcommittees, has been working on the development of a revision to the Standard to embrace comments, recommendations, and experiences since the original 2002 edition, and to supplement the loading requirements of *ASCE/SEI 7-10, Minimum Design Loads for Buildings and Other Structures*, since the latter does not include requirements for loads during construction. The environmental loads provisions of this ASCE/SEI 37-14 have been aligned with those of ASCE/SEI 7-10 and adjusted for the duration of the construction period.

Final committee balloting was completed, and public comments solicited and resolved in mid-2014.

ACKNOWLEDGMENTS

Design Loads on Structures during Construction, Standard ASCE/SEI 37-14, was developed over a period of several years by the Design Loads on Structures during Construction Standard Committee of the Codes and Standards Activities Division (CSAD) of the Structural Engineering Institute (SEI), and of the Codes and Standards Activities Committee (CSAC) of the American Society of Civil Engineers (ASCE). This 2014 edition was prepared by six subcommittees of the Design Loads on Structures during Construction Standard Committee under the leadership of the following individuals:

Robert T. Ratay, Ph.D., P.E., F.SEI, F.ASCE, *Chairman*
Rubin M. Zallen, P.E., F.ASCE, Chapter 1.0, General
John F. Duntemann, P.E., S.E., M.ASCE, Chapter 2.0, Loads and Load Combinations
Cris Subrizi, P.E., M.ASCE, Chapter 3.0, Dead and Live Loads
John S. Deerkoski, P.E., S.E., P.P., F.ASCE, and Alan Fisher, P.E., M.ASCE, Chapter 4.0, Construction Loads
Vincent Tirolo Jr., P.E., M.ASCE, Chapter 5.0, Lateral Earth Pressure
James G. Soules, P.E., S.E., P.Eng., SECB, F.SEI, F.ASCE, Chapter 6.0, Environmental Loads

The particularly active long-term participation and valuable contribution of the following members, in addition to the chair, the five subcommittee chairs, and two co-chairs, is acknowledged: James R. Harris, Gilliam S. Harris, Donald Dusenberry, and David W. Johnston.

ASCE acknowledges the work of the Design Loads on Structures during Construction Standard Committee of the CSAD of SEI and of the CSAC of ASCE. The Standard Committee comprises individuals from many backgrounds including design, analysis, research, consulting engineering, construction, education, government, and private practice.

Voting members of the 42-member Standard Committee are:

Robert T. Ratay, Ph.D., P.E., F.SEI, F.ASCE, Chair
Cosema E. Crawford, P.E., M.ASCE
John S. Deerkoski, P.E., S.E., P.P., F.ASCE, Chapter 4 Subcommittee Co-Chair
John F. Duntemann, P.E., S.E., M.ASCE, Chapter 2 Subcommittee Chair
Donald Dusenberry, P.E., F.ASCE, F.SEI
Alan Fisher, P.E., M.ASCE, Chapter 4 Subcommittee Co-Chair
Noel J. Gardner
David H. Glabe, P.E., M.ASCE
Ram A. Goel, P.E., F.ASCE
Allan H. Gold, P.E., S.E., R.A., F.ASCE
Vijayalakshmi Gopalrao, Aff.M.ASCE
Dennis W. Graber, P.E., L.S., M.ASCE
Gilliam S. Harris, P.E., F.ASCE
James R. Harris, Ph.D., P.E., F.SEI, M.ASCE
Philip T. Hodge, P.E., M.ASCE
William P. Jacobs V, P.E., S.E., M.ASCE
David W. Johnston, Ph.D., P.E., F.ASCE
Roger S. Johnston, P.E., M.ASCE
Robert C. Krueger, P.E., M.ASCE
David G. Kurtanich Sr., P.E., M.ASCE
Jim E. Lapping, P.E., M.ASCE
John V. Loscheider, P.E., M.ASCE
Robert G. McCracken
Bob Glenn McCullouch, P.E., M.ASCE
Thomas J. Meany Jr., P.E., F.ASCE
Brian Medcalf, P.E., M.ASCE
Joel Moskowitz, P.E., D.GE, M.ASCE
Joe N. Nunnery, P.E., M.ASCE
Dai H. Oh, P.E., M.ASCE
David B. Peraza, P.E., M.ASCE
Max L. Porter, Ph.D., P.E., F.SEI, Dist.M.ASCE
Emil Simiu, Ph.D., P.E., F.EMI, F.ASCE
John M. Simpson, P.E., M.ASCE
James G. Soules, P.E., S.E., P.Eng., SECB, F.SEI, F.ASCE, Chapter 6 Subcommittee Chair
Cris Subrizi, P.E., M.ASCE, Chapter 3 Subcommittee Chair
Bruce A. Suprenant, P.E., M.ASCE
Raymond H. R. Tide, Ph.D., P.E., F.ASCE
Vincent Tirolo Jr., P.E., M.ASCE, Chapter 5 Subcommittee Chair
Ronald W. Welch, Ph.D., P.E., M.ASCE
Michael A. West, P.E., F.SEI, F.ASCE
Terry K. Wright, P.E., M.ASCE
Rubin M. Zallen, P.E., F.ASCE, Chapter 1 Subcommittee Chair

CHAPTER 1
GENERAL

STANDARD

1.1 PURPOSE

The purpose of this standard is to provide minimum design load requirements during construction of buildings and other structures.

1.2 SCOPE

This standard addresses partially completed structures, temporary structures, and temporary supports used during construction. The loads specified herein are suitable for use either with strength design [such as ultimate strength design (USD) or load and resistance factor design (LRFD)] or with allowable stress design (ASD). The loads are equally applicable to all conventional construction materials.

This standard does not specify the party responsible for the design of temporary structures or temporary supports, or for the temporary use of incomplete structures. This standard also does not specify the party responsible for on-site supervision of the construction of temporary structures or temporary supports, or for the use of incomplete structures.

1.3 BASIC REQUIREMENTS

1.3.1 Structural Integrity Partially completed structures and temporary structures shall possess sufficient structural integrity, under all stages of construction, to remain stable and resist the loads specified herein.

Stability of the incomplete structure and the possibility of progressive collapse shall be considered.

COMMENTARY

C1.1 PURPOSE

The construction loads, load combinations, and load factors contained herein account for the often short duration of loading, and for the variability of temporary loads. Many elements of the completed structure that are relied upon implicitly to provide strength, stiffness, stability, or continuity are sometimes not present during construction.

The requirements in this standard complement those in ASCE/SEI 7-10 (ASCE/SEI 2010).

C1.2 SCOPE

In this standard, loads and load factors are based on probability where sufficient probabilistic information is available. Where there is insufficient probabilistic information, loads and load factors are established by experience and engineering judgment.

Safety factors for ASD and capacity reduction factors for strength design are not given in this standard. They are given or implied in the structural design standards for the various structural materials.

This standard is not intended to account for loads caused by gross negligence or error.

This standard is intended for use by engineers knowledgeable in the performance of structures.

Responsibilities for the design of temporary structures and temporary supports, for the design for temporary use of partially completed structures, and for supervision of site activities to control loads on structures are contractual matters that should be resolved among the parties involved in the construction of a structure. This standard is not intended to establish the responsible parties for any of those activities.

The requirements contained herein are not intended to adversely affect the selection of a particular construction material or type of construction.

C1.3 BASIC REQUIREMENTS

C1.3.1 Structural Integrity Structural integrity should be provided by sequencing the construction so as to avoid creating vulnerable partially completed portions of the structure; by completing the permanent system that supports lateral loads as the dependent portions of the structure are erected; by avoiding conditions that result in loads that exceed the capacity of structural elements and their connections; or otherwise by providing temporary supports.

In many cases, stronger members and connections than originally designed for the permanent structure will have to be provided to support construction loads.

Many structures experience higher wind loading on the open structure than on the completed, enclosed structure,

STANDARD

COMMENTARY

which may require temporary bracing in addition to the permanent lateral load-resisting system.

During erection of a structure, the permanent lateral load-resisting system is generally not complete. Also, other elements of the structural system that are essential to the overall performance of the structure may not be in place or may only be partially secured. As such, the structure may be vulnerable to severe and widespread damage should a single, local failure or mishap occur.

1.3.2 Serviceability The effects of construction loads or conditions shall not adversely affect the serviceability or performance of the completed structure.

C1.3.2 Serviceability An example of adverse effect on serviceability is excessive permanent deformation.

1.3.3 Types of Loads All loads in this standard—dead, live, construction, environmental, and lateral earth pressure—shall be considered. In addition, forces resulting from the interaction between the partially completed structure and temporary supporting or bracing structures shall be considered.

C1.3.3 Types of Loads The loads in this standard may be different from those used in the design of the completed structures.

Although this standard is limited to the determination of the types of loads listed in this section, the consideration of other loads and effects may also be appropriate for specific conditions or when specified by the authority having jurisdiction.

1.3.4 Construction The effects on the loads created by the methods and sequencing (scheduling) of the construction during the progressive stages of the work shall be considered.

1.3.4.1 Construction Methods Limiting Loads Construction methods that limit loads by allowing controlled damage or controlled failure of temporary construction (including parts of the structure being built) when subject to extreme environmental loads are permitted subject to a mitigation plan approved by the authority having jurisdiction.

C1.3.4.1 Construction Methods Limiting Loads An example of limiting loads is the design of a cofferdam to a specified tidal height, allowing it to flood under higher water levels.

The mitigation plan should provide for monitoring of the environmental effects that could cause the extreme loads; an evacuation plan; protection of personnel on the construction site; protection of the public; and protection of the structures being built and adjacent structures.

1.3.5 Analysis Load effects on incomplete structures, on temporary structures, and on their respective individual components shall be determined by accepted methods of structural and geotechnical analysis, taking into account second-order effects.

1.4 ALTERNATE CRITERIA FOR DESIGN LOADS DURING CONSTRUCTION

The use of other authoritative documents shall be permitted for design loads during construction, for a specific material or method of construction, when acceptable to the authority having jurisdiction.

C1.4 ALTERNATE CRITERIA FOR DESIGN LOADS DURING CONSTRUCTION

Authoritative documents such as specifications and reports of research or studies applicable to specific materials or methods of construction have been developed and are recognized and used extensively. Examples include AASHTO *Standard Specifications for Highway Bridges* (AASHTO 2002), AASHTO *LRFD Bridge Design Specifications* (AASHTO 2007), AASHTO *Guide Design Specifications for Bridge Temporary Works* (AASHTO 1995), CALTRANS *Memo to Designers 20-2* (CALTRANS 2011), CALTRANS *Memo to Designers 15-14* (CALTRANS 2010), the Mason Contractors Association of America's

STANDARD

COMMENTARY

Standard Practice for Bracing Masonry Walls Under Construction (MCAA 2012), and the National Concrete Masonry Association's TEK 3-4B, *Bracing Concrete Masonry Walls During Construction* (NCMA 2005).

The user is cautioned that the bases of loads and the corresponding load factors in authoritative documents permitted under Section 1.4 may be different from those in this standard. For example, wind velocity in this standard (based on ASCE 7-10) is the average of the 3-s gust given for a return period that yields loads at the strength level; the corresponding load factor is 1.0. The wind velocities in other standards are determined for shorter return periods or for the fastest mile of wind, yielding loads at the allowable stress level; the corresponding load factors for those loads are 1.3 or 1.6. Also, gust factors for wind velocities measured by the 3-s gust are different from gust factors for wind velocities measured by the fastest mile of wind. The user should not mix the load from another document, the basis of which is different from the load in this standard, with a load factor from this standard without making an appropriate adjustment to the magnitude of the load.

REFERENCES

American Association of State Highway and Transportation Officials (AASHTO). (1995 with 2008 Interim Revisions). *Guide design specifications for bridge temporary works*, Washington, DC.

AASHTO. (2002). *Standard specifications for highway bridges*, 17th Ed., Washington, DC.

AASHTO. (2007). *LRFD bridge design specifications*, 4th Ed., Washington, DC.

ASCE/SEI. (2010). "Minimum design loads for buildings and other structures." *ASCE/SEI 7-10, Including Supplement No. 1*, Reston, VA.

California Department of Transportation (CALTRANS). (2010). "Loads for temporary highway structures." *Memo to Designers 15-14, December 2010, with Attachment 1*, Sacramento, CA.

CALTRANS. (2011). "Site seismicity for temporary bridges and staged construction." *Memo to Designers 20-2, May 2011*, Sacramento, CA.

Mason Contractors Association of America (MCAA). (2012). *Standard practice for bracing masonry walls under construction*, Lombard, IL.

National Concrete Masonry Association (NCMA). (2005). "Bracing concrete masonry walls during construction." *TEK 3-4B*, Herndon, VA.

CHAPTER 2
LOADS AND LOAD COMBINATIONS

STANDARD

2.1 LOADS SPECIFIED

Structures within the scope of this standard shall resist the effects of the following loads and combinations thereof:

Dead and live loads: see Chapter 3
D = dead load; and
L = live load.

Construction loads: see Chapter 4
Weight of temporary structures
C_D = construction dead load.

Material loads
C_{FML} = fixed material load; and
C_{VML} = variable material load.

Construction loads
C_P = personnel and equipment loads;
C_H = horizontal construction load;
C_F = erection and fitting forces;
C_R = equipment reactions; and
C_C = lateral pressure of concrete.

Lateral earth pressures: see Chapter 5
C_{EH} = lateral earth pressures.

Environmental loads: see Chapter 6
W = wind;
T = thermal loads;
S = snow loads;
E = earthquake;
R = rain; and
I = ice.

The specified loads are nominal loads which are intended to be suitable for use in either conventional allowable stress design (ASD) or load and resistance factor design (LRFD), provided that appropriate load factors and combinations are used.

2.2 LOAD COMBINATIONS AND LOAD FACTORS FOR STRENGTH DESIGN

Specified loads shall be combined according to the principles in this section to obtain the maximum design load effects for members and systems.

2.2.1 Additive Combinations The applicable load combinations shall be identified and evaluated in accordance with this section. The total design load for each combination shall be the sum of the factored dead and/or material loads present, the variable load(s) at their maximum

COMMENTARY

C2.1 LOADS SPECIFIED

In this section the loads are defined only by name and symbol. The complete definition and specification of each load is in the referenced sections.

The occupancy or use live load, L, is different from the construction loads C_{FML}, C_{VML}, and C_P. The occupancy or use live load and the construction loads are mutually exclusive, in that either an area is under construction, in which case the construction loads are applied, or an area is not under construction but is being used for a certain occupancy, in which case the appropriate live load, L, for that occupancy is applied. During construction, in most situations, there will be no live load applied to the structure.

Additional construction and environmental loads may include (and be accounted for in the load combinations) such items as differential settlement, prestressing, shrinkage, rib shortening, stream flow pressure, buoyancy, and others as appropriate.

C2.2 LOAD COMBINATIONS AND LOAD FACTORS FOR STRENGTH DESIGN

The selection of load factors is intended to be compatible with ASCE/SEI 7-10 (ASCE/SEI 2010). Because little independent research has been done, the load factor 2.0 is suggested for those loads that may vary substantially, or about which we have little information. AASHTO provides additional guidance for load combinations and factors for use with bridge temporary works (AASHTO 2002, 2007, 2012a).

C2.2.1 Additive Combinations The loads suggested herein for consideration in load combinations are not all-inclusive; therefore, their selection will require judgment in many situations. Design should be based on the load combination causing the most unfavorable effect. In some

STANDARD

values, and the other uncorrelated loads at their arbitrary point-in-time (APT) values. Correlated variable loads, such as vertical and horizontal construction loads, shall be taken to have their maximum values occurring simultaneously. The generalized form of the load combinations (U) can be written as:

Combined design load =
dead and/or material loads
+ loads at their maximum values
+ loads at their arbitrary point-in-time reduced values.

$$U = \sum_k c_{D,k} D_{n,k} + \sum_i c_{\max,i} Q_{n,i} + \sum_j c_{APT,j} Q_{n,j} \quad (2\text{-}1)$$

where c_D = dead load factor; $c_{\max}$ = load factor for the maximum value of variable load; c_{APT} = load factor for the arbitrary point-in-time value of variable load; D_n = nominal dead or construction material load; Q_n = nominal variable load; k = all dead and construction material loads; i = all loads occurring at maximum value; and j = all relevant simultaneously occurring variable loads at APT values.

Whenever different variable loads are correlated, such as horizontal and vertical loads from the same source or operation, the same load factor, $c_{\max}$ or c_{APT}, shall be used in Eq. (2-1) for these loads.

2.2.2 Load Factors The minimum load factors for use with strength design shall be as follows:

Load	Load factor ($c_{\max}$)	Arbitrary point-in-time load factor (c_{APT})
D	0.9 (when counteracting wind or seismic loads)	–
	1.4 (when combined with only construction and material loads)	–
	1.2 (for all other combinations)	
L	1.6	0.5
C_D	0.9 (when counteracting wind or seismic loads)	–
	1.4 (when combined with only construction and material loads)	–
	1.2 (for all other combinations)	–
C_{FML}	1.2	–
C_{VML}	1.4	By analysis
C_P	1.6	0.5
C_C	1.2 (full head)	–
	1.6 (otherwise)	–
C_{EH}	1.6	–
C_H	1.6	0.5

COMMENTARY

cases, this may occur when one or more loads are not applied simultaneously. Furthermore, the critical load effect may result from the application of one or more loads on only part of the structure. Finally, concentrated loads may be applied in place of or in addition to the assumed uniformly distributed loads.

Load combinations should be considered based on the specific type of construction and procedures. Consideration should be given to construction loads which may be mutually exclusive, strongly correlated, or occur with such a low probability that they may effectively be neglected.

The unfactored loads to be used in the combinations herein are the nominal loads in Chapters 3 through 6 of this standard. The concept of using maximum and APT loads and corresponding load factors is consistent with ASCE/SEI 7-10. Here, in addition to the dead load (which is assumed to be permanent), one or more of the variable loads takes on its maximum value, while the other variable loads occurring simultaneously assume APT values (i.e., those values measured at any instant of time). This is consistent with the way loads actually combine in situations in which strength limit states are approached. The nominal loads in Chapters 3 through 6 are substantially in excess of the APT values. Rather than providing both a maximum and an APT nominal load value for each load type, load factors of less than unity are provided for APT loads.

C2.2.2 Load Factors The load factors provided herein are intended to reflect the relative uncertainty in the particular action. This uncertainty can arise from (1) inherent or natural variability, (2) range of applications, and (3) possibilities for misuse or error. It may therefore be reasonable to make certain modifications to load factors in the presence or absence of additional information.

For example, a lower load factor is specified for conditions of full fluid head when designing for lateral pressure of concrete because this condition suggests less uncertainty than partial (unknown) fluid head.

Factors on heavy-equipment reactions are for maximum load values only. Because this is considered in combinations only when it is actually *present*, no APT factor is provided. Further, the load factor is much lower (1.6) when the equipment is rated such that the reactions are specified by the manufacturer or are otherwise known. Also, in the event any equipment is used that generates dynamic loads (e.g., pumps, unbalanced rotors), the load effect must be determined separately first and then multiplied by a factor of 1.3.

Environmental loads are considered in a similar way as in ASCE/SEI 7-10. However, the following differences for environmental loads during construction must be kept in mind: (1) modifications to the design load values for the possibility of a reduced exposure period is appropriate, (2) certain loads may be disregarded for most practical purposes because of the generally very short reference period associated with typical construction projects, and (3) certain loads in combinations may effectively be

STANDARD | COMMENTARY

Load	Load factor (c_{max})	Arbitrary point-in-time load factor (c_{APT})
C_F	2.0	By analysis
C_R	2.0 (unrated)	0
	1.6 (rated)	0
W	1.0	0.3
T	1.2	–
S	1.6	0.5
E	1.0	–
R	1.6	–
I	1.6	–

ignored because of the practice to shut down work sites during these events (e.g., snow and wind, snow and certain equipment forces, extreme winds, and personnel loads). Regional and project-specific conditions should be considered when deciding which combinations of environmental and structural loads to use.

The designer should be aware that temporary structures used repeatedly are subject to abuse and loss of capacity, and that capacity reduction factors (φ factors) may need to be lower than those used for ordinary strength design to compensate for this loss of capacity.

2.2.3 Combinations Using Strength Design Except where applicable codes and standards provide otherwise, structures and their components shall be designed so that their strength exceeds the effects of factored loads in the following combinations:

$$1.4D + 1.4C_D + 1.2C_{FML} + 1.4C_{VML} \quad (2\text{-}2)$$

$$1.2D + 1.2C_D + 1.2C_{FML} + 1.4C_{VML} + 1.6L \quad (2\text{-}3)$$

$$1.2D + 1.2C_D + 1.2C_{FML} + 1.4C_{VML} + 1.6C_P + 1.6C_H + 0.5L \quad (2\text{-}4)$$

$$1.2D + 1.2C_D + 1.2C_{FML} + 1.4C_{VML} + 1.0W + 0.5C_P + 0.5L \quad (2\text{-}5)$$

$$1.2D + 1.2C_D + 1.2C_{FML} + 1.4C_{VML} + 1.0E + 0.5C_P + 0.5L \quad (2\text{-}6)$$

$$0.9D + 0.9C_D + (1.0W \text{ or } 1.0E) \quad (2\text{-}7)$$

where D = dead load in place at the stage of construction being considered, L = live load, which may be less than or greater than the final live load, and W = wind load computed using the design velocity reduction per Section 6.2.1.

The most unfavorable effects from both wind and earthquake loads shall be considered, where appropriate, but they need not be assumed to act simultaneously. Similarly, C_H need not be assumed to act simultaneously with wind or seismic loads. Lateral earth pressure, environmental loads, and other construction loads shall be considered when applicable, using the load factors in Section 2.2.2. Consideration of these other loads will require the use of load combinations in addition to the basic combinations listed above.

C2.2.3 Combinations, Using Strength Design OSHA (2013) requires that "Each scaffold and scaffold component shall be capable of supporting, without failure, its own weight and at least 4 times the maximum intended load applied or transmitted to it." ANSI (2011) has a similar requirement. In order to satisfy this OSHA criterion, the load factor for personnel and equipment load, C_P, fixed material load, C_{FML}, and variable material load, C_{VML}, should be 4.0 and the load factor for construction dead load, C_D, should be 1.0. However, assuming $C_D = 1.0$ may be unconservative in cases where the dead load is high relative to the total load. Also, φ factors used with these load factors should be 1.0.

The load factor on wind load in Eqs. (2-5) and (2-7) has been reduced from 1.6 in ASCE 7-95 (the edition of ASCE 7 referenced in the previous edition of ASCE 37) to 1.0 in ASCE/SEI 7-10. Therefore, ASCE 37-14 has been revised accordingly. This reduction is necessary because of the change in the specification of the design wind speed in Chapter 26 of ASCE/SEI 7-10. As explained in the Commentary to ASCE/SEI 7-10, the wind speed is now mapped at much longer return periods than in the previous editions of ASCE 7.

2.2.4 Counteracting Combinations Where the effect of one load is partially or wholly resisted by another load, the factor on the resisting load shall be taken as zero for variable loads and 0.90 for controlled loads.

C2.2.4 Counteracting Combinations Section 2.2.4 applies to conventional structural systems not subject to OSHA regulations. Counteracting combinations for scaffold systems are defined by OSHA regulations.

STANDARD

A controlled load is a material load that is placed in a specific location to counteract the effect of another load.

2.3 ALLOWABLE STRESS DESIGN

Specified loads shall be combined according to this section to obtain the maximum design load effects for members and systems.

Increases in allowable stress shall not be used with the loads or combinations given in this standard unless it can be demonstrated that such an increase is justified by structural behavior caused by rate or duration of load.

2.3.1 Additive Combinations When using load values provided in this standard for ASD, sufficient additive load combinations shall be considered to obtain the maximum design load effects for members and systems.

The following basic combinations shall be investigated as a minimum:

$$D + C_D + C_{FML} + C_{VML} + L \quad (2\text{-}8)$$

$$D + C_D + C_{FML} + C_{VML} + C_P + C_H + L \quad (2\text{-}9)$$

$$D + C_D + C_{FML} + C_{VML} + 0.6W + C_P + L \quad (2\text{-}10)$$

$$D + C_D + C_{FML} + C_{VML} + 0.7E + C_P + L \quad (2\text{-}11)$$

$$0.6D + C_D + (0.6W \text{ or } 0.7E) \quad (2\text{-}12)$$

where D = dead load in place at the stage of construction being considered, L = live load, which may be less than or greater than the final live load, and W = wind load computed using the design velocity factor where appropriate per Section 6.2.1.

The most unfavorable effects from both wind and earthquake loads shall be considered where appropriate, but they need not be considered simultaneously. Similarly, C_H need not be assumed to act simultaneously with wind or seismic loads. Other construction loads which shall be considered if applicable are defined in Section 2.1.

2.3.2 Load Reduction When structural effects due to two or more variable loads, in combination with dead load, are investigated in load combinations of Section 2.3.1, the combined effects shall comply with both of the following requirements: (1) the combined effects due to the two or more variable loads multiplied by 0.75 plus effects due to dead loads shall not be less than the effects from the load

COMMENTARY

C2.3 ALLOWABLE STRESS DESIGN

Designers are cautioned against mixing allowable stress design (ASD) and load and resistance factor design (LRFD) load combinations.

OSHA (1993) requires that "Each scaffold and scaffold component shall be capable of supporting, without failure, its own weight and at least 4 times the maximum intended load applied or transmitted to it." ANSI (2011) has a similar requirement. Allowable stress design ordinarily provides for safety factors somewhat less than 2; thus, in order to satisfy the OSHA criteria, the superimposed design loads effectively need to be more than doubled.

The designer should be aware that temporary structures used repeatedly are subject to abuse and loss of capacity, and that safety factors may need to be higher than those used for ordinary ASD to compensate for this loss of capacity.

C2.3.1 Additive Combinations As with the strength provisions described previously, the possible loads shown in this document are not all-inclusive, and designers will need to exercise judgment in selecting the appropriate loads to be considered in combination for the given construction situation. Design should be based on the load combination causing the most unfavorable effect. In some cases, this may occur when one or more loads are not acting. In other cases, the governing load effect may result from loads not applied over the entire element being designed. As described previously, designers also need to be aware of construction loads which may be mutually exclusive, strongly correlated, or which occur with such a low probability that they may be neglected.

Similar to the load factors in Section 2.2.3, the wind load in Eqs. (2-10) and (2-12) has been reduced from 1.0W to 0.6W to reflect the changes from ASCE 7-95 (the edition of ASCE 7 referenced in the previous edition of ASCE 37) to ASCE/SEI 7-10. Similarly, the 0.6 factor has been applied to the dead load in Eq. (2-12) to provide comparable reliability between strength design and ASD. More discussion on this revision can be found in the Commentary to ASCE/SEI 7-10.

STANDARD

combination of the dead load plus the load producing the largest effects, and (2) the allowable stress shall not be increased to account for these combinations.

2.3.3 Counteracting Combinations Where the effect of one load is partially or wholly resisted by another load, the factor on the resisting load shall be taken as zero for variable loads and 0.6 for controlled loads.

2.4 BRIDGES

Load combinations for design loads on bridges during construction shall conform to generally accepted design codes or specifications for such work.

COMMENTARY

C2.4 BRIDGES

Generally accepted design codes and specifications for design loads on bridges during construction include the following: the AASHTO *LRFD Bridge Construction Specification* (AASHTO 2012b), the AREMA *Manual for Railway Engineering* (AREMA 2005), and the AASHTO *Guide Design Specification for Bridge Temporary Works* (AASHTO 2012a).

The first edition of the AASHTO *LRFD Bridge Construction Specifications* (AASHTO 2012b) was adopted in 1998. With the adoption of the AASHTO *LRFD Bridge Construction Specifications*, AASHTO moved the Division II: Construction provisions that appeared in the *Standard Specifications for Highway Bridge Structures* to the AASHTO *LRFD Bridge Construction Specifications*.

The AASHTO *LRFD Bridge Construction Specifications* requires that the design of temporary works be in accordance with the AASHTO *LRFD Bridge Design Specifications* or the *Guide Design Specifications for Bridge Temporary Works* unless another recognized specification is accepted by the engineer (presumably the state bridge engineer or governing authority) (AASHTO 2002). The AASHTO *LRFD Bridge Construction Specifications* also notes that the design of access scaffolding is subject to OSHA regulations.

CALTRANS and ASCE provide guidance to designers regarding loads for temporary highway structures to supplement the AASHTO specifications (ASCE 1981; CALTRANS 2010).

REFERENCES

American Association of State Highway and Transportation Officials (AASHTO). (2002). *AASHTO standard specifications for highway bridges*, 17th Ed., Washington, DC.

AASHTO. (2007). *LRFD bridge design specifications*, 4th Ed., Washington, DC.

AASHTO. (2012a). *Guide design specifications for bridge temporary works*, 2nd Ed., Washington, DC.

AASHTO. (2012b). *LRFD bridge construction specifications*, 3rd Ed., Washington, DC.

American National Standards Institute (ANSI). (2011). "Scaffolding: Safety requirements for construction and demolition operations." *ANSI A10.8-2011*, Washington, DC.

American Railway Engineering and Maintenance-of-Way Association (AREMA). (2005). *Manual for railway engineering*, Landover, MD.

ASCE. (1981). "Recommended design loads for bridges." *J. Struct. Div.*, 107(ST7), 1161–1213.

STANDARD

COMMENTARY

ASCE/SEI. (2010). "Minimum design loads for buildings and other structures." *ASCE/SEI 7-10, Including Supplement No. 1*, Reston, VA.

California Department of Transportation (CALTRANS). (2010). "Loads for temporary highway structures." *Memo to Designers 15-14, December 2010, with Attachment 1*, Sacramento, CA.

Occupational Safety and Health Administration (OSHA). (2013). *Occupational safety and health standards (Parts 1910 and 1926)*, Department of Labor, Washington, DC. (OSHA regulations are continually reviewed and revised. Sections other than those referenced here may also apply.)

CHAPTER 3
DEAD AND LIVE LOADS

STANDARD

3.1 DEAD LOADS

The dead load, D, for the purposes of this standard, is the weight of the permanent construction in-place at the particular time in the construction sequence that is under consideration. The dead load includes all construction in-place that is temporarily shored or braced. It includes construction for which the primary structural system is complete, but which is being used to support construction materials and construction equipment. The weights of scaffolding, shoring, concrete forms, runways for construction equipment, temporary bridges, and other temporary structures are not included; these loads are considered to be construction dead load, C_D, as defined in Section 4.1.1.

The weight of the permanent construction in-place includes all nonstructural loads such as cladding, partitions, ceilings, and railings that are expected to be in place at the particular time being considered.

3.2 LIVE LOADS

The live load, L, is the load produced by the use or occupancy of a structure that is under construction. These loads may be imposed on construction in-place, on partially demolished structures, and on temporary structures. The live load, L, may vary at different stages of construction.

For bridge structures and other transportation structures, live load shall include impact, longitudinal forces from vehicles, centrifugal forces from vehicles, and wind loads on vehicles, as applicable.

COMMENTARY

C3.1 DEAD LOADS

The contractor usually controls the sequence of construction and thus controls what loads will be on the structure at the various construction stages. The design of temporary shoring and bracing must include these dead loads as well as the temporary loads described in Chapters 3, 4, 5, and 6, as applicable.

Tables of common construction dead loads are provided in ASCE 7 *Minimum Design Loads for Buildings and Other Structures*, in building codes, and in various engineering handbooks.

C3.2 LIVE LOADS

Live load may be present in a structure that is being remodeled, underpinned, or otherwise repaired, replaced, or demolished in stages.

The live loads during construction may be different from the live loads applied on the completed structure. For example, during reconstruction of a bridge designed for trucks, a lane may be restricted to cars, resulting in a lower live load. On the other hand, temporary overcrowding of a completed section of a building would warrant an increase in live load. Reduction of the live load from the final design value shall not be made unless the use of the facility is strictly monitored and enforced.

The partially completed structure, or partially demolished structure, should expose the occupants or users to no greater risk than inherent in the codes and standards of practice that pertain to the completed structure.

Ideally, the design drawings will identify live loads to be applied during construction, if applicable.

CHAPTER 4
CONSTRUCTION LOADS

STANDARD

4.1 GENERAL REQUIREMENTS

The provisions of Chapter 4 shall be used to define the construction loads for the design of both temporary structures and permanent structures subject to loads during construction. These loads are to be combined with other applicable loads per the requirements of Chapter 2.

Alternatively, other authoritative documents that address construction loads for a specific material or method of construction shall be permitted to be used in accordance with Section 1.4.

4.1.1 Definitions

Construction loads: Those loads imposed on a partially completed or temporary structure during and as a result of the construction process. Construction loads include, but are not limited to, materials, personnel, and equipment imposed on the temporary or permanent structure during the construction process.

Construction dead load, C_D: The dead load of temporary structures that are in place at the stage of construction being considered. The dead load of the permanent structure, either partially complete or complete, is not included in C_D; the dead load of the permanent structure is defined as dead load, D, in Section 3.1.

Individual personnel load: A concentrated load of 250 lb (1.11 kN) that includes the weight of one person plus equipment carried by the person or equipment that can be readily picked up by a single person without assistance.

Working surfaces: Floors, decks, or platforms of temporary or partially completed structures which are or are expected to be subjected to construction loads during construction.

COMMENTARY

C4.1 GENERAL REQUIREMENTS

The loads for some temporary structures, such as those that retain lateral pressures of earth, are not defined in Chapter 4; refer to Chapter 5 for lateral pressures of earth.

An example of an authoritative document which meets the intention of Section 1.4 would be an edition of ACI 347 (ACI 2004) that is issued after SEI/ASCE 37-02 (SEI/ASCE 2002). See Section C1.4 for other examples of documents that are permitted under Section 1.4.

Other documents traditionally used for the determination of construction design loads include the following:

AISC 303-10 (AISC 2010a) and *Specifications for Structural Steel Buildings* 360-10 (AISC 2010b), ANSI A 10.9-1997 (R2004), ANSI's *Safety Requirements for Steel Erection* A10.13-2001 (ANSI 2001) and *Scaffolding: Safety Requirements for Construction and Demolition Operations* A10.8-2011 (ANSI 2011), PCI's *Erector's Manual: Standards and Guidelines for the Erection of Precast Concrete Products* MNL-127-99 (PCI 1999), the CALTRANS *Falsework Manual* (CALTRANS 2001), the Federal Highway Administration's *Guide Design Specification for Bridge Temporary Works* FHWA-R.D.-93-032 (FHWA 1993), the New York City Transit Authority's *Field Design Standards* (New York City Transit Authority Engineering & Construction Dept. 1986), the Metal Building Manufacturers Association's *Metal Building Systems Manual* (MBMA 2006), Chapter 19 in the *Handbook of Temporary Structures in Construction* (Marshall 1996), and "Discussion on the Guide to Formwork for Concrete" (ACI 1989).

STANDARD

4.2 MATERIAL LOADS

The material dead loads consist of two categories:

1. Fixed material loads (FML), and
2. Variable material loads (VML).

Fixed material loads (FML) are loads from materials that are fixed in magnitude. Variable material loads (VML) are loads from materials that vary in magnitude during the construction process. If the local magnitude of a material load varies during the construction process, then that load must be considered as a variable material load.

4.2.1 Concrete Load The weight of concrete placed in a form for the permanent structure is a material load. When the concrete gains sufficient strength so that the formwork, shoring, and reshoring are not required for its support, the concrete becomes a dead load.

COMMENTARY

C4.2 MATERIAL LOADS

This section separates material dead loads into two categories, fixed material loads (FML) and variable material loads (VML). Fixed material loads (FML) and variable material loads (VML) are separated to permit the use of an appropriate load factor for each category in strength design. This approach recognizes the difference in the variability of the load between the two categories.

This section addresses the loads from materials and is not intended to apply to equipment loads. Personnel and equipment loads are considered separately in Section 4.3.

Material loads may be either distributed or concentrated loads. The designer must consider the pattern of uniformly distributed loads and the location of concentrated loads that create the most severe strength and/or serviceability condition.

The designer must determine whether the superimposed material load during the construction process is essentially fixed in magnitude, or is variable, or can be adjusted during the construction process. For example, the load from formwork becomes a FML once it is installed, whereas the load created by the concrete during fresh concrete placement is considered a VML. The load due to concrete placement is considered a variable material load because fresh concrete can be piled higher than the finished thickness of the slab.

The distinction between a FML and a VML is not location or position on the structure; rather, it is the variability of the loading magnitude.

The stockpiling of any material is considered a VML (e.g., scaffold, forms, rebar, metal deck, barrels, drywall, ceiling tile, roofing materials). Some materials, such as scaffold or forms, are considered VML when stockpiled but may be considered FML when placed in their final end-use position. Engineering judgment must be used to determine whether the stockpiled material should be considered as a uniformly distributed or a concentrated load.

Stockpiled materials should be positioned on the structure to minimize the effects of early loading on the serviceability or performance of the completed structure.

Careful consideration should be given to the placement of stockpiled materials on early-age concrete structures. Early loading of low-strength concrete has been shown to increase long-term deflection (Fu and Gardner 1986; Sbarounis 1984; Yamamoto 1982). It is recommended that materials be stockpiled at columns, avoiding placement in the middle of long spans. This practice serves to decrease long-term deflections of reinforced concrete beams and to deter lateral buckling of unsupported steel beams.

For a list of the proper weights for different building materials, the designer should consult ASCE/SEI 7-10 (ASCE/SEI 2010), or other specified or recognized sources.

STANDARD

4.2.2 **Materials Contained in Equipment** Materials being lifted by or contained in equipment are part of the equipment load, not a material load. Once such material has been discharged from the equipment, it becomes a material load.

4.3 PERSONNEL AND EQUIPMENT LOAD, C_P

4.3.1 General Personnel and equipment loads shall be considered in the analysis or design of a partially completed or temporary structure. The design or analysis of the structure shall be governed by uniformly distributed and/or a concentrated personnel and equipment load, whichever creates the most severe strength and/or serviceability condition. The governing load shall be placed in the pattern or location that creates the most severe strength and/or serviceability condition.

The personnel and equipment loads used in the design or analysis of a partially completed or temporary structure shall be the maximum loads that are likely to be created during the sequence of construction.

4.3.2 Uniformly Distributed Loads Uniform loads shall be selected to result in forces and moments that envelope the forces and moments that would result from the application of concentrated loads that could occur and which are not separately considered.

4.3.3 Concentrated Loads The personnel and equipment concentrated loads shall be the maximum loads expected in the construction process, but shall be no less than those given in Table 4-1. The concentrated load shall be located to produce the maximum strength and/or serviceability conditions in the structural members. The designer shall consider each category of minimum concentrated personnel and equipment load that is likely to occur during the construction process.

Concentrated loads from equipment shall be determined in accordance with Section 4.6.

For temporary structures that are used for public traffic, the structure shall be designed in accordance with the AASHTO bridge design specifications (AASHTO 2002, 2010), or the construction equipment that will use the structure in accordance with the requirements of Section 4.6, whichever gives the more critical effects.

COMMENTARY

C4.2.2 Materials Contained in Equipment The equipment reactions should include the effects of the material being lifted or contained therein. See Section 4.6.

C4.3.2 Uniformly Distributed Loads Construction loads, except for material loads, will rarely be distributed uniformly. However, design for equivalent uniformly distributed loads is a long-standing practice that has stood the test of time. The designer determining the effects of loads during construction must select a uniform load that will adequately capture the effects of anticipated construction loads. Section 4.8.1.1 presents a tabulation of traditional minimum uniformly distributed loads that include personnel, equipment, and material in transit or staging.

C4.3.3 Concentrated Loads The designer must make an important decision in choosing the concentrated load category that properly fits the construction process for the project.

Concentrated loads from equipment are a serious concern. The type of equipment to be used for each construction operation, its location (on or off the structure), and its loading must be considered. Loads for different types of construction equipment have been tabulated (Jahren 1996; Caterpillar 2008). See also Sections C4.6.1 and C4.6.2 for precautions in using tabulated data.

Individual Personnel Load, defined in Section 4.1.1, is defined in ANSI 10.8 (ANSI 2011) as 200 lb (0.89 kN) per person plus 50 lb (0.22 kN) of equipment per person; however, the totals of the loads are the same.

Wheeled vehicles, both manually operated and powered, may require a more rigorous analysis similar to AASHTO. The factors that may have to be investigated are:

- Pneumatic tire pressure;
- Spacing of adjacent tires;
- Axle load;
- Number of axles;
- Spacing of axles; and
- Gross vehicle weight.

STANDARD

Table 4-1. Minimum Concentrated Personnel and Equipment Loads

Action	Minimum Load[a] [lb (kN)]	Area of Load Application [in. × in. (mm × mm)]
[b]Each person	250 (1.11)	12 ×12 (300 × 300)
Wheel of manually powered vehicle	500 (2.22)	Load divided by tire pressure[c]
Wheel of powered equipment	2,000 (8.90)	Load divided by tire pressure[c]

[a]Use actual loads where they are larger than tabulated here.
[b]The spacings of the 250-lb concentrated loads need not be less than 18 in. (457 mm) c. to c.
[c]For hard rubber tires, distribute load over an area 1 in. (25 mm) × the width of the tire.

4.3.4 Impact Loads The concentrated loads specified in Table 4-1 include adequate allowance for ordinary impact conditions. Provision shall be made in the structural design for loads that involve predictable unusual vibration and impact forces.

4.4 HORIZONTAL CONSTRUCTION LOAD, C_H

One of the following horizontal load criteria, where appropriate, shall be applied to temporary or partially complete structures as a minimum horizontal loading, whichever gives the greatest structural effects in the direction under consideration:

1. For wheeled vehicles transporting materials, 20% for a single vehicle or 10% for two or more vehicles of the fully loaded vehicle weight. Said force shall be applied in any direction of possible travel, at the running surface,
2. For equipment reactions as described in Section 4.6, the calculated or rated horizontal loads, whichever are the greater,
3. 50 lb/person (0.22 kN/person), applied at the level of the platform in any direction, or
4. Two percent of the total vertical load. This load shall be applied in any direction and shall be spatially distributed in proportion to the mass. This load need not be applied concurrently with wind or seismic load.

This provision shall not be considered as a substitute for the analysis of environmental loads.

COMMENTARY

In some instances the authority having jurisdiction may require that provisions be made for a specified load.

Many specifications require temporary or permanent structures to be designed for a uniform load and/or a concentrated load. If the source of the concentrated load can be clearly identified, such as wheel loads, axle loads, pallet loads, or equipment reactions, that specific load should be distributed as determined by its source.

Problems arise in determining the distribution areas of unidentified, but specified, loads. To determine the distribution area for an unidentified concentrated load, assume that the load will be generated by the densest material normally available on a construction site. That material is arbitrarily chosen to be concrete at 150 pcf (2,400 kg/m^3), and of a cubic shape.

The specified concentrated load in Table 4-1 is assumed to be the total load, including dynamic forces.

The concentrated loads required herein are not intended for protection against an accident involving falling objects, such as when a beam, a length of reinforcing steel, or a piece of equipment falls several stories.

C4.3.4 Impact Loads The designer determining loads during construction is not expected to anticipate the effects of poor workmanship such as concrete being discharged from a bucket from excessive heights above the formwork. A concrete bucket that hits the forms is an impact load that would be considered accidental and would not fall within the scope of this provision.

C4.4 HORIZONTAL CONSTRUCTION LOAD, C_H

Forces necessary for member stability are determined during analysis of the structure, and as such are not specified by this standard.

The intent of this provision is to provide a minimum lateral load-resistance mechanism and a minimum lateral stiffness in all temporary or partially complete structures. Unavoidable eccentricities might cause vertical superimposed loads to produce some horizontal loading. Also, horizontal loads can be created from personnel and equipment operations.

The designer should be aware that the actual horizontal loads may exceed the minimum specified in this section, particularly if more than one construction activity is being conducted at the same time.

The 50 lb/person (0.22 kN/person) load in Criterion 3 represents a conservative estimate of the lateral force that could be generated from the activities of personnel.

Criterion 4 is intended to provide a minimum lateral load resistance and to assure lateral stability, for the structure as-a-whole, during construction. Generally, it is not expected that this criterion will result in forces during construction that exceed the capacity of the permanent lateral load-resisting system of a structure below the level where the permanent lateral load-resisting system has been completed; however, the permanent lateral load-resisting system needs to be checked for this criterion.

STANDARD | COMMENTARY

Wind and other phenomena that produce horizontal loads must be considered separately from the requirements of this section, as specifically provided in Criterion 4.

4.5 ERECTION AND FITTING FORCES, C_F

Forces due to erection (e.g., alignment, fitting, bolting, bracing, guying) shall be considered.

4.5.1 Anchorage of Steel Columns The column attachment to the foundation shall be designed to resist a minimum eccentric gravity load of 300 lb (1.33 kN) located 18 in. (457 mm) from the extreme outer face of the column in each direction at the top of the column shaft.

4.5.2 Steel Column Splices Column splices shall be designed to resist a minimum eccentric gravity load of 300 lb (1.33 kN) located 18 in. (457 mm) from the extreme outer face of the column in each direction of the top of the column shaft.

C4.5 ERECTION AND FITTING FORCES, C_F

This provision applies to all types of structures but more specifically to the erection of components typical of steel, metal, timber, and precast structures.

C4.5.1 Anchorage of Steel Columns This requirement is consistent with a requirement of OSHA Subpart R, Steel Erection (OSHA 1993).

See OSHA Subpart R for clarification on the difference between columns and posts.

C4.5.2 Steel Column Splices This requirement is consistent with a requirement of OSHA Subpart R, Steel Erection (OSHA 1993).

4.6 EQUIPMENT REACTIONS, C_R

The reactions from equipment, with due consideration of all loading conditions, shall be used in the design of the temporary or partially completed structure. The equipment reactions shall include the full weight of the equipment operating at its maximum rated load in conjunction with any applicable environmental loads unless the use is restricted and revised reactions are developed.

4.6.1 General The structure shall be designed to safely support the full weight of the equipment and associated worst-case load effects due to its operation. The design shall include the consideration of support deflections or movements, out-of-level supports, vertical misalignment, and environmental loads on the equipment.

4.6.2 Rated Equipment The minimum equipment loads for design shall be those provided by the equipment manufacturer or supplier.

Unless loaders, such as front-end loaders or fork lifts, are intentionally restricted from tipping on one axle, the loader self-weight plus tipping load shall be applied to the front axle.

C4.6 EQUIPMENT REACTIONS, C_R

Rated equipment is that for which reactions are given by the equipment manufacturer or supplier. For nonrated equipment, the designer is to determine the reactions by analysis. Shapiro et al. (1999) provides examples of calculations for reactions from lifting or hoisting equipment that include assessments of environmental loads.

C4.6.1 General In addition to utilizing a piece of equipment at less than its maximum operating capacity, there may be hybrid situations such as the use of a crane with rear outriggers placed over a major support member of the structure. At this point, the crane may reach out to its maximum operating radius to make a pick and then boom in, thus substantially reducing the outrigger reactions when the crane swings to a new position where it would deposit or pick up the load. In this case, maximum outrigger loads are apparently developed over the rear of the crane and lesser loads are developed over the other outrigger pads which could be placed on lighter structural members. This is a common practice where the outrigger support members are not adequate to sustain the full or maximum rated capacity outrigger reactions.

As a result of a shift of the center of gravity, vehicle axle loads and crane outrigger or support reactions may be greatest in the absence of payload or pick. The worst-case condition controls (loaded or unloaded).

C4.6.2 Rated Equipment Care should be exercised when using the tabulated values for equipment, such as loaders, from references or from any manufacturer's data. Axle load distributions at maximum load do assume that all of the axles are touching the ground and with a certain load distribution. Unless special precautions are taken, such as limiting bucket size and floor or deck obstacles, it

STANDARD

The designer shall verify the basis of the rating and the rated reactions given by the equipment supplier. If the basis of the rating is different from the conditions under which the equipment will be used, then the more severe reactions shall be used in design.

4.6.3 Nonrated Equipment The equipment loads for nonrated equipment shall be determined by analysis or testing.

4.6.4 Impact The reaction of equipment shall be increased by 30% to allow for impact, unless other values (either larger or smaller) are recommended by the manufacturer, or are required by the authority having jurisdiction, or are justified by analysis.

4.7 FORM PRESSURE

4.7.1 Form Pressure Unless the conditions of Section 4.7.1.1 or 4.7.1.2 are met, formwork shall be designed for the lateral pressure of the newly placed concrete given by Eq. (4-1). Minimum values given for other pressure formulas do not apply to Eq. (4-1).

$$C_C = wh \tag{4-1}$$

$$C_{C\,SI} = \rho g h_{SI} \tag{4-1 SI}$$

where C_C ($C_{C\,SI}$) = lateral pressure, psf (kPa); w = unit weight of fresh concrete, pcf; ρ = density of concrete (kg/m^3); g = gravitational constant, 0.00981 kN/kg; and h (h_{SI}) = depth of fluid or plastic concrete from top of placement to point of consideration, ft (m).

For columns or other forms that may be filled rapidly before any stiffening of the concrete takes place, h shall be taken as the full height of the form, or the distance between horizontal construction joints when more than one placement of concrete is to be made. When working with mixtures using newly introduced admixtures that increase set time or increase slump characteristics, such as self-consolidating concrete, Eq. (4-1) shall be used until the effect on formwork pressure is understood by measurement.

4.7.1.1 For concrete having a slump of 7 in. (175 mm) or less, and placed with normal internal vibration to a depth of 4 ft (1.2 m) or less, formwork may be designed for a lateral pressure as follows. F_C and F_W are defined in Tables 4-2 and 4-3.

For columns:

$$C_C = F_C F_W (150 + 9000\ R/T) \tag{4-2}$$

$$C_{C\,SI} = F_C F_W \left[7.2 + \frac{785\ R_{SI}}{T_{SI} + 17.8} \right] \tag{4-2 SI}$$

COMMENTARY

is a quite frequent occurrence that the loaders, in attempting to pick materials for transport, will either catch an element of the deck or try to pick more than their rated load. In this instance, the entire vehicle picks up and pivots about its front axle. This load could create axle and wheel loads more than 30% greater than the manufacturer's rated wheel load.

C4.7 FORM PRESSURE

C4.7.1 Form Pressure The lateral pressure formulas are adopted from ACI 347-04 (ACI 2004; Hurd 2005). Equation (4-1) assumes a fully liquid head and normally can be applied without restriction. However, there are exceptions. Caution must be taken when using external vibration or concrete made with shrinkage compensating cement. In these situations, pressures in excess of equivalent hydrostatic may occur.

The designer must consider the uplift caused by the vertical component of the normal pressure of freshly placed concrete on inward-sloping forms.

The SI version of the lateral pressure formulas is adopted from ACI 347-04.

Slip form pressures are not covered in this document.

See also ACI (1989) and Barnes and Johnston (2003).

C4.7.1.1 Under the limitations listed, the formwork may be designed for a maximum lateral pressure, as provided in Eqs. (4-2)–(4-4), that is less than the full hydrostatic head.

Where any of the limitations are not met, the lateral pressure must be taken as provided in Eq. (4-1).

STANDARD

COMMENTARY

with a minimum of 600 F_W psf (30 F_W kPa), but in no case greater than wh ($\rho g h_{SI}$).

For walls with a rate of placement of less than 7 ft/h (2.1 m/h) and a placement height not exceeding 14 ft/h (4.2 m/h):

$$C_C = F_C F_W (150 + 9000\,R/T) \quad (4\text{-}3)$$

$$C_{C\,SI} = F_C F_W \left[7.2 + \frac{785\,R_{SI}}{T_{SI} + 17.8}\right] \quad (4\text{-}3\ SI)$$

with a minimum of 600 F_W psf (30 F_W kPa), but in no case greater than wh ($\rho g h_{SI}$).

For walls with a rate of placement of less than 7 ft/h (2.1 m/h) where placement height exceeds 14 ft/h (4.2 m/h) and for all walls with a placement rate of 7 to 15 ft/h (2.1 to 4.5 m/h):

$$C_C = F_C F_W [150 + 43{,}400/T + 2800\,R/T] \quad (4\text{-}4)$$

$$C_{C\,SI} = F_C F_W \left[7.2 + \frac{244\,R_{SI}}{T_{SI} + 17.8} + \frac{1156}{T_{SI} + 17.8}\right] \quad (4\text{-}4\ SI)$$

with a minimum of 600 F_W psf (30 F_W kPa), but in no case greater than wh ($\rho g h_{SI}$),

where:

R (R_{SI}) = rate of placement, ft/h (m/h)
T (T_{SI}) = temperature of concrete in the form, °F (°C)
F_C = chemistry factor per Table 4-2
F_W = unit weight factor per Table 4-3

Table 4-2. Chemistry Factor F_C

Cement type of blend	F_C
Types I, II, and III without retarders[a]	1.0
Types I, II, and III with a retarder[a]	1.2
Other types or blends containing less than 70% slag or 40% fly ash without retarders[a]	1.2
Other types or blends containing less than 70% slag or 40% fly ash with a retarder[a]	1.4
Blends containing more than 70% slag or 40% fly ash	1.4

[a]Retarders include any admixture, such as a retarder, retarding water reducer, retarding midrange water-reducing admixture, or high-range water-reducing admixture (superplastizer), that delays setting of concrete.

STANDARD

Table 4-3. Unit Weight Factor F_W

Inch-pound version	
Unit weight of concrete	F_W
Less than 140 lb/ft^3	$0.5[1 + (w/145\,lb/ft^3)]$
But not less than 0.80	
140–150 lb/ft^3	1.0
More than 150 lb/ft^3	$w/145\,lb/ft^3$
SI version	
Density of concrete	F_W
Less than 2,240 kg/m^3	$0.5[1 + (w/2{,}320\,kg/m^3)]$
But not less than 0.80	
2,240–2,400 kg/m^3	1.0
More than 2,400 kg/m^3	$w/2{,}320\,kg/m^3$

4.7.1.2 If concrete is pumped from the base of the form, the form shall be designed for full hydrostatic head of concrete, $C_C = wh$, plus a minimum allowance of 25% for pump surge pressure.

4.7.2 (Reserved for future use)

4.7.3 Shoring Loads Where shores are required to support the load of newly placed concrete, these shores shall be maintained until the concrete has gained enough strength to support applicable dead and construction loads. Where shoring is continuous over several floors, the calculated loads on these shores shall be cumulative unless and until the shores have been released and reset to allow the slab in question to carry its own dead and construction loads. Such release shall not occur until the concrete is capable of carrying its own dead load.

4.8 APPLICATION OF LOADS

4.8.1 Combined Loads The design construction load shall include the critical combination of personnel, equipment, and material loads.

COMMENTARY

C4.7.1.2 In certain instances, pressures may be as high as the face pressure of the pump piston.

C4.7.3 Shoring Loads ACI 347.2R-05 (ACI 2005) provides recommendations for analysis of shoring loads and the distribution of these loads to the structure in multistory concrete construction. Release of shoring/reshoring may also require adherence to timing limitations arising out of project deflection criteria, if specified.

C4.8 APPLICATION OF LOADS

Construction loads depend very much on the specific planning and processes of construction. This section includes rules for applying and combining the various loads, as well as traditional minimums for several common construction processes.

C4.8.1 Combined Loads The combination of the various forms of construction loads, materials, personnel, and equipment is an important step in engineering for construction, requiring careful application of professional judgment.

STANDARD

4.8.1.1 Working Surfaces Structures supporting working surfaces as defined in Section 4.1 shall be designed for the combined material, personnel, equipment, and other applicable construction loads.

Where the construction operation fits the definition in Table 4-4, the designer is permitted to design for the tabulated uniform loads as the vertical load from the combination of personnel, equipment, and material in transit or staging. Where the construction operation does not fit the definitions in Table 4-4, the design shall be for the actual loads. Concentrated loads shall be considered separately.

Table 4-4. Classes of Working Surfaces for Combined Uniformly Distributed Loads

Operational Class	Uniform Load[a] [psf (kN/m²)]
Very Light Duty: sparsely populated with personnel, hand tools, very small amounts of construction materials.	20 (0.96)
[b]Light Duty: sparsely populated with personnel, hand-operated equipment, staging of materials for lightweight construction.	25 (1.20)
[b]Medium Duty: concentrations of personnel, staging of materials for average construction.	50 (2.40)
[b]Heavy Duty: material placement by motorized buggies, staging of materials for heavy construction.	75 (3.59)

[a]Loads do not include dead load, D; construction dead load, C_D, or fixed material loads, C_{FML}.
[b]OSHA categories.

COMMENTARY

C4.8.1.1 Working Surfaces It is traditional to design many working surfaces for a uniformly distributed load that is meant to include all construction loads, except for materials in their final position.

Temporary structures have often been designed, advertised, and specified by the Light, Medium, and Heavy Duty ratings given in Table 4-4. This standard also applies to partially completed structures, and the same terminology is adopted. Different styles of construction and different segments of the construction industry have different traditions for design loads on partially completed structures during construction, and this section of the standard is an attempt to unify the industry on a common basis.

Examples of construction operations that have traditionally been designed for the loads given in the table are:

Very Light Duty:
- Roofing, reroofing, excepting situations with stockpiles of ballast
- Access catwalks
- Painting, caulking
- Maintenance using hand tools

Light Duty:
- Light frame construction
- Concrete transport and placement by hose and concrete finishing with hand tools

Medium Duty:
- Concrete transport and placement by buckets, chutes, or handcarts
- Concrete finishing using motorized screeds
- Masonry construction with tile or hollow lightweight concrete units
- Structural steel erection or concrete reinforcing steel placement

Heavy Duty:
- Concrete transport and placement using motorized buggies
- Masonry of brick or heavyweight concrete units
- Material storage

Conflicts between provisions of this section and those in ASCE 3-91 (ASCE 1991a) and ASCE 9-91 (ASCE 1991b) are acknowledged.

Examples of working surfaces that do not fall under Table 4-4 are:

- Roofs where design is controlled by building code live load or snow loads that are less than the values in Table 4-4; and
- Attic or hung ceilings which provide access for maintenance, installation of utilities, and emergency services such as firefighters.

These working surfaces must be addressed in accordance with Sections 4.8.1.1 and 4.8.4.

STANDARD

4.8.1.2 Specification of Temporary Structures Where temporary structures are specified by load name, the names of the load class and the magnitude of design loads shall be as given in Table 4-4.

4.8.2 Pattern Loading The designer shall consider patterns of construction loads when such loadings produce more demanding effects than does application of the full intensity of the construction load over the entire structure.

4.8.3 Reduction in Construction Loads

4.8.3.1 Material Loads No reduction is allowed for fixed or variable material loads, except to the extent that small amounts of material in transit or staging are included in uniformly distributed personnel, equipment, and material loads, such as those in Table 4-4.

4.8.3.2 Personnel and Equipment Loads When justified by an analysis of the construction operations, members having an influence area of 400 ft^2 (37.16 m^2) or more shall be permitted to be designed for a reduced uniformly distributed personnel and equipment load determined by applying the following formula:

$$C_P = L_o\left(0.25 + 15/\sqrt{A_I}\right) \quad (4\text{-}5)$$

$$C_{P\,SI} = L_o\left(0.25 + 4.57/\sqrt{A_I}\right) \quad (4\text{-}5\ \text{SI})$$

where C_P ($C_{P\,SI}$) = reduced design uniformly distributed personnel and equipment load per ft^2 (m^2) of area supported by the member; L_o = unreduced uniformly distributed personnel and equipment design load per ft^2 (m^2) of area supported by member; and A_I = influence area, in ft^2 (m^2). The influence area A_I is normally four times the tributary area for a column, two times the tributary area for a beam, and equal to the panel area for a two-way slab.

The reduced uniformly distributed personnel and equipment design load, regardless of influence area, shall not be less than 50% of the unreduced design load for members supporting one level, or 40% of the unreduced design load for members supporting more than one level, except that where the uniformly distributed personnel and equipment load is 25 psf (1.2 kN/m^2) or less, the reduced load shall not be less than 60% of the unreduced design load, unless justified by an analysis of the construction operations.

4.8.3.3 Personnel and Equipment Loads on Sloping Roofs A reduction in gravity construction loads for personnel and equipment on a roof is also permitted based upon the slope of the roof. The reduction factor, *R*, is:

$$R = 1.2 - 0.05F \quad (4\text{-}6)$$

COMMENTARY

C4.8.1.2 Specification of Temporary Structures This requirement will encourage uniformity in terminology for capacity of scaffolds and similar structures.

C4.8.2 Pattern Loading ASCE/SEI 7-10 describes other possible conditions of designing members or floors for partial loading. Pattern loadings on continuous beams and floors of single-story and multistory frames can produce the highest positive and negative moments and shears in specific elements. The designer should consider the possibility that partial loading of a structure, or full loading of a partially completed structure, will create demands that exceed those calculated assuming that the full structure is fully loaded.

C4.8.3.2 Personnel and Equipment Loads Uniformly distributed loads are a convenient substitute for computing the combined effect of several concentrated loads. As such, they are generally calibrated to a particular area. For smaller areas, the concentrated loads control structural design. The nature of transient concentrated loads, such as personnel and equipment, is that their spacing is not uniform; thus, for areas larger than the calibration area, the uniform load may be unnecessarily conservative.

A construction load reduction based on influence is reasonable. There is a lack of data from construction projects. Without specific information, the derivation of a new reduction equation was not warranted, and therefore a commonly used live load reduction procedure (ASCE-SEI 7-10) has been used for this document.

Care shall be exercised because many construction loads are actual, not statistical, loads. If actual loads are anticipated over the entire area, no reduction should be taken.

For load restrictions see Section C4.8.4.

ASCE/SEI 7-10 allows for reduced live loads of 12 psf (0.57 kN/m^2). Model building codes have the same roof loading. On the surface, this is a violation of OSHA minimum loading of 25 psf (1.2 kN/m^2) (OSHA 1993); however, the OSHA requirement is for temporary platforms and traditionally has not been applied to completed or partially completed roof structures. See Section 4.8.4.

C4.8.3.3 Personnel and Equipment Loads on Sloping Roofs For consistency, the reduction in roof personnel and equipment loads also follows ASCE/SEI 7-10. The detail of application is somewhat different, but the limits are essentially the same.

STANDARD

where F = slope of the roof expressed in in./ft (in SI system, F = 0.12 × slope of the roof expressed in percentage points). R need not exceed 1.0 and R shall not be less than 0.6. This reduction may be combined by multiplication with the reduction based upon area, but the reduced load shall not be less than 60% of the basic unreduced load.

4.8.4 Restriction of Loads The following working surfaces shall have their use and access restricted by posting of the permitted loads and load conditions or by operational control by the entity which has jurisdiction over their use:

1. Scaffolds with working surfaces of 40 ft^2 (3.72 m^2) or less shall be rated for the number of Individual Personnel Loads that they can support, and the working surfaces shall be restricted accordingly. When designing, the Individual Personnel Loads shall be placed in such locations as to maximize their effects on the structural members of the scaffold; however, they need not be spaced closer than 2 ft-0 in. (0.61 m) on center.
2. Working surfaces designed for superimposed uniform loads of 25 psf (1.20 kN/m^2) or less shall be rated for both their superimposed uniform load capacity and the number and location of the Individual Personnel Loads that they can support. These working surfaces shall be restricted accordingly.
3. Working surfaces designed for loads less than what could reasonably be expected to be placed thereon shall be restricted to the design loads.

COMMENTARY

C4.8.4 Restriction of Loads Posting, restricting, or otherwise limiting construction loads is consistent with building codes, AASHTO, OSHA, the scaffolding industry, and ANSI requirements. This issue should not be confused with reduction of design live load based on contributory area. Load reduction, based on influence area, is addressed in Section 4.8.3.2.

The posting or load restriction can be accomplished by physical barriers that direct the traffic on a bridge deck or parking structure, or barriers on a floor system to restrict access to wheeled vehicles, storage of materials, or personnel.

It is not uncommon to have relatively large work platforms or scaffolds, for example, a 100 × 100-ft (30.5 × 30.5-m) work platform for renovation of the structural steel roof of a building or members of a bridge. As a platform for personnel and equipment, the heavy-duty 75 psf (3.59 kN/m^2) rating is appropriate to design the deck system. However, recognizing that the work crew may consist of several personnel working in a localized area, using the 75 psf (3.59 kN/m^2) rating is inappropriate simultaneously on the entire platform, and does not reflect the true operating use of the scaffold or platform. The hanger or support system for the platform could be designed for the maximum load developed by the limited number of personnel on the platform clustering around one support to create the greatest load at that support point. The platform would be clearly posted or rated, as with scaffolds in accordance with the ANSI 10.8 (ANSI 2011) maximum number of occupants. Failure to do this could result in a loading on the structure from which the scaffold is hanging or supported from substantially exceeding the design load of that structure.

There are a myriad of lightweight platforms and scaffolds which are only intended to support one to three persons, their small tools, and incidental materials. An example is small hanging platforms below steel beams ("floats") used extensively in structural steel erection. The capacities of these scaffolds are controlled by the number of Individual Personnel Loads for which they are designed. Restriction of use is necessary to prevent overloading.

REFERENCES

American Association of State Highway and Transportation Officials (AASHTO). (2002). *AASHTO standard specifications for highway bridges*, 17th Ed., Washington, DC.

AASHTO. (2010). *LRFD bridge design specifications*, 6th Ed., Washington, DC.

American Concrete Institute (ACI). (1989). "Discussion on the guide to formwork for concrete." *ACI Struct. J.*, 86(3), 320–323.

STANDARD

COMMENTARY

ACI. (2004). "Guide to formwork for concrete." *347-04*, Farmington Hills, MI.

ACI. (2005). "Guide for shoring/reshoring of concrete multistory buildings." *347.2R-05*, Farmington Hills, MI.

American Institute of Steel Construction (AISC). (2010a). "Erection." Section 7, *Code of standard practice for steel buildings and bridges, AISC 303-10*, Chicago.

AISC. (2010b). "Erection." Section M4, *Specifications for structural steel buildings, AISC 360-10*, Chicago.

American National Standards Institute (ANSI). (1997). "Safety requirements for concrete and masonry work: American national standard for construction and demolition operations." *ANSI A10.9-1997 (R2004)*, Washington, DC.

ANSI. (2001). "Safety requirements for steel erection." *ANSI A10.13-2001*, Washington, DC.

ANSI. (2011). "Scaffolding: Safety requirements for construction and demolition operations." *ANSI A10.8-2011*, Washington, DC.

ASCE. (1991a). "Standard for the structural design and construction of composite slabs." *ASCE 3-91*, Reston, VA.

ASCE. (1991b). "Standard practice for construction and inspection of composite slabs." *ASCE 9-91*, Reston, VA.

ASCE/SEI. (2010). "Minimum design loads for buildings and other structures." *ASCE/SEI 7-10, Including Supplement No. 1*, Reston, VA.

Barnes, J. M., and Johnston, D. W. (2003). "Fresh concrete lateral pressure on formwork." *Proc., ASCE-CI Construction Research Congress 2003*, Reston, VA.

California Department of Transportation (CALTRANS). (2001). *Falsework manual*, Rev. 32, Office of Structure Construction, Sacramento, CA.

Caterpillar performance handbook. (2008). Caterpillar, Inc., Sanford, NC.

Federal Highway Administration (FHWA). (1993). "Guide design specification for bridge temporary works." *FHWA-R.D.-93-032*, Washington, DC.

Fu, H. C., and Gardner, N. J. (1986). "Effect of high early-age construction loads on the long term behavior of slab structures." *Properties of concrete at early ages, ACI SP-95*, J. F. Young, ed., American Concrete Institute, Farmington Hills, MI.

Hurd, M. K. (2005). *Formwork for concrete*, 7th Ed., American Concrete Institute, Farmington Hills, MI.

Jahren, C. T. (1996). "Loads created by construction equipment." Chapter 6, *Handbook of temporary structures in construction*, 2nd Ed., R. T. Ratay, ed., McGraw-Hill, New York.

Marshall, R. (1996). "Bracing and guying for stability." Chapter 19, *Handbook of temporary structures in construction*, R. T. Ratay, ed., McGraw-Hill, New York.

Metal Building Manufacturers Association (MBMA). (2006). *Metal building systems manual*, Cleveland, OH.

New York City Transit Authority Engineering & Construction Dept. (1986). *Field design standards*, New York.

Occupational Safety and Health Administration (OSHA). (1993). *Safety and health regulations for construction, Subpart R.* Code of Federal Regulations, Part 1926, Title 29, Chapter XVII, Department of Labor, Washington, DC. (OSHA regulations are continually reviewed and revised. Sections other than those referenced here may also apply.)

Prestressed Concrete Institute (PCI). (1999). "Erector's manual: Standards and guidelines for the erection of precast concrete products." *MNL-127-99*, Chicago.

SEI/ASCE. (2002). "Design loads on structures during construction." *SEI/ASCE 37-02*, ASCE, Reston, VA.

Sbarounis, J. A. (1984). "Multi-story flat plate buildings: Effect of construction loads on long-term deflections." *Concrete Int.*, 6(4): 62–70.

Shapiro, H. I., Shapiro, J. P., and Shapiro, L. K. (1999). *Cranes and derricks*, 3rd Ed., McGraw-Hill, New York.

Yamamoto, T. (1982). "Long-term deflections of reinforced concrete slabs subjected to overloading at an early age." *Proc., Int. Conf. on Concrete at Early Ages*, Paris, April 6–8, 1982.

CHAPTER 5
LATERAL EARTH PRESSURE

STANDARD

5.1 DEFINITION

For the purpose of this standard, lateral earth pressure, C_{EH}, is defined as the horizontal or nearly horizontal resultant of forces per unit area created by soil and water on a vertical or nearly vertical plane of a structure.

5.2 DETERMINATION OF LATERAL EARTH PRESSURE

Design values of lateral earth pressures and their distribution shall be determined by the use of credible and reliable methods in accordance with accepted engineering practice. One test of credibility of a method of earth pressure determination shall be its publication in one or more generally accepted references on geotechnical engineering.

Site-specific conditions shall be considered in the selection of method(s) of calculation, and site-specific data shall be used for the critical factors in the calculations.

Distinction shall be made among active, at-rest, and passive earth pressures as influenced by the direction and magnitude of movement or deformation of the structure under load.

Laboratory or field instrumentation, observations, and measurements shall be permissible bases for determination of earth pressures.

COMMENTARY

C5.2 DETERMINATION OF LATERAL EARTH PRESSURE

The magnitude and distribution of soil pressures on both permanent and temporary structures during construction depend on a multitude of factors. Their determination for design should be performed by an engineer with an adequate knowledge of soil mechanics, an understanding of structural behavior, and a familiarity with the construction procedures.

In order for lateral soil pressures to occur, the soil behind the temporary excavation support system must undergo strain (AASHTO 2008b). The magnitude of this strain is not only a function of the in situ ground conditions, but also wall stiffness and construction staging. The control of ground movement behind ground support systems is often the controlling factor in design.

Innovative methods of temporary excavation supports are continually developing. The application of existing methods of earth pressure analyses should be done with caution and should, whenever possible, rely on recorded field performance in similar ground conditions.

The following technical documents are readily obtainable and are considered to be credible references for the calculation of lateral earth pressures: AASHTO, 2008a; AASHTO, 2002; ASCE, 1997; Caltrans, 2011; Canadian Geotechnical Society, 2006; NAVFAC DM-7.2; FHWA-RD-75-128, 129, 130; FHWA-IF-99-015.

REFERENCES

American Association of State Highway and Transportation Officials (AASHTO). (2008a). *Construction handbook for bridge temporary work*. Washington, DC.

AASHTO. (2008b). *Guide design specifications for bridge temporary works*. Washington, DC.

AASHTO. (1996). "Retaining walls." Section 5, *Standard specifications for highway bridges*, 17th Ed., Washington, DC.

ASCE. (1997). "Guidelines of engineering practice for braced and tied-back excavations." *Geotechnical special publication No. 74*, Reston, VA.

California Department of Transportation. (2011). *Trenching and shoring manual*, Sacramento, CA.

Canadian Geotechnical Society. (2006). *Canadian foundation engineering manual*, 4th Ed., Richmond, BC, Canada.

Department of the Navy, Naval Facilities Engineering Command. (1982). "Foundations and earth structures." *Design Manual 7.2, NAVFAC DM-7.2*, Washington, DC.

Goldberg, D. T., Jaworsky, W. E., and Gordon, M. D. (1976). "Lateral support systems and underpinning." *Reps. FHWA-RD-75-128, 129, 130*, Federal Highway Administration, Washington, DC.

STANDARD

COMMENTARY

Sabatini, P. J., Pass, D. G., and Bachus, R. C. (1999). "Ground anchors and anchored systems." *Geotechnical Engineering Circular No. 4, Publication No. FHWA-IF-99-015*, Federal Highway Administration, Washington, DC.

CHAPTER 6
ENVIRONMENTAL LOADS

STANDARD

The basic reference for computation of environmental loads is the 2010 edition of ASCE 7 (ASCE/SEI 2010). The requirements of ASCE/SEI 7-10 shall apply except as modified herein.

COMMENTARY

This section deals with special issues of construction and temporary structures for which the basic procedures of ASCE 7-10 (ASCE/SEI 2010) are to be modified.

The environmental loads in this chapter are reduced from those in ASCE 7-10 in recognition of the anticipated lifespan of temporary structures and temporary configurations of structures under construction. Reduction to the safety of individuals is not the intent of the committee.

Reductions of loads to the levels stated in this standard are appropriate when loading situations can be managed through safety protocols that limit access to hazardous locations when loadings exceed those used for temporary designs, and when loadings, including environmental loadings, can be limited (e.g., by timely snow removal) proactively. The knowledge and training of personnel in control of construction sites, the visible nature of construction elements, and the processes on construction sites are key components of protocols necessary to control of risk to personnel and property on the construction site. Risks to personnel and property adjacent to the construction site also warrant attention.

Structures and structural configurations designed for the loads specified in this section will likely sustain damage, should they be exposed to the environmental events that form the basis for the loads specified in ASCE/SEI 7-10 (ASCE/SEI 2010). Due to the low probability of the extreme environmental event occurring combined with the short duration of the construction of the structures and structural configurations, the risk of damage is considered acceptable. This is particularly acknowledged for severe earthquake loadings, which cannot be anticipated in advance, but have a very low probability of impacting any particular construction site during a construction period.

The load reductions used in ASCE 37 are statistically based on the assumption that configurations designed to exist for specific construction periods will, indeed, be modified or removed before those construction periods end. It is not appropriate to use this standard to justify reduced loads by artificially subdividing a continuous construction operation into parts that each appear to create shorter construction periods and, therefore, lower design loads. It also is not appropriate to attempt to extend the life of the design of an interim condition that had a specific original construction period by proclaiming at any point in its life that the expected remaining duration into the future does not exceed the original construction period.

Standards and other documents applicable to specific materials or methods of construction have been developed and are recognized and used extensively. Examples include AASHTO *Standard Specifications for Highway Bridges* (AASHTO 2002), CALTRANS *Memo to Designers 20-2* (CALTRANS 1989), CALTRANS *Memo to Designers 20-12* (CALTRANS 2003), CALTRANS *Memo to Designers 15-14* (CALTRANS 2004b), and the Masonry Contractors Association of America's *Standard Practice for*

STANDARD | COMMENTARY

Bracing Masonry Walls Under Construction (MCAA 2001). In addition, the designer needs to comply with all applicable laws, ordinances, and regulations, including those of the Occupational Safety and Health Administration. In some cases, these regulatory requirements may be more restrictive than those contained in ASCE 37 and its cited references.

6.1 RISK CATEGORY

Unless otherwise required by the authority having jurisdiction, the risk category, as defined in ASCE/SEI 7-10, shall be taken as Risk Category II for all environmental loads during construction, regardless of the risk category assigned for the design of the completed structure.

C6.1 RISK CATEGORY

During construction, the primary occupancy of a building is by construction personnel. As such, the risk of loss of human life is comparable to that for Risk Category II buildings as defined in ASCE 7-10. Circumstances in which the engineer may consider a higher risk category, or in which some authorities have required such consideration, include construction work immediately adjacent to essential facilities in which a construction failure would imperil operation of the essential facility.

6.2 WIND

Except as modified herein, wind loads shall be calculated in accordance with procedures in ASCE/SEI 7-10.

Design wind pressures shall be based on design wind speeds calculated in accordance with Section 6.2.1. The minimum wind pressure of 16 psf (0.77 kN/m^2) specified by ASCE/SEI 7-10 need not be applied.

C6.2 WIND

Structures in intermediate stages of construction are frequently susceptible to the effects of wind.

Information and guidance have been lacking in the United States on the selection of wind speeds and force coefficients on structures during construction (Ratay 1987). Limited research and development have been performed for the purpose of this standard (Boggs and Peterka 1992; Rosowsky 1995).

If local conditions so dictate, and for certain hazardous construction operations, it might be appropriate to apply a minimum strength design level wind pressure, such as 16 psf (0.77 kN/m^2), to the design.

Section 6.2 has been revised to refer to ASCE/SEI 7-10. The 2010 edition of ASCE 7 (ASCE/SEI 2010) uses a new basis for the calculation of wind speeds. The wind speeds used in ASCE/SEI 7-10 are strength design-level wind speeds. Therefore, the wind speeds referred to in Section 6.2, in addition to the load combinations found in Chapter 2, have been revised to reflect this change.

6.2.1 Design Wind Speed The design wind speed shall be taken as the following factor times the basic wind speed in ASCE/SEI 7-10, except as required in Section 6.2.1.1.1.

Construction Period	Factor
Less than six weeks	0.75
From six weeks to one year	0.8
From one to two years	0.85
From two to five years	0.9

C6.2.1 Design Wind Speed Wind speeds are reduced from requirements for permanent structures, consistent with the philosophy explained in Section C6.0. The quantitative method used to achieve this objective is that the wind load should have the same likelihood of being exceeded in the construction period as the permanent structure design wind does in a 50-year period. The reduced construction period speed factors have been developed to achieve this objective (Boggs and Peterka 1992; Rosowsky 1995).

Factors for construction periods less than one year are developed based on judgment, because statistical analyses of seasonal wind variations have not been performed for all regions. Local wind speed data should be consulted when using these factors.

STANDARD

COMMENTARY

6.2.1.1 Construction Period The construction period shall be taken as the time interval from first erection to structural completion of each independent structural system, including installation of cladding.

For construction periods of less than six weeks, factors of less than 0.75 shall be permitted if justified by a statistical analysis of local wind data for the season during which the subject construction conditions will exist.

6.2.1.1.1 Construction Period in Hurricane-Prone Areas For construction between November 1 and June 30 (outside of the hurricane season), the basic wind speed of 115 mph (51 m/s) shall be permitted for structures sited near the Gulf Coast and Eastern Seaboard, where the ASCE/SEI 7-10 specified basic wind speed exceeds 115 mph (51 m/s) (3 second gust) (hurricane-prone areas). The 115 mph (51 m/s) wind speed is permitted to be reduced by the factors in Section 6.2.1 only for a construction period between November 1 and June 30. If the construction period shifts into the period between July 1 and October 31, the design shall be reviewed and modified, as appropriate, to conform to the requirements shown below for a construction period between July 1 and October 31.

Between July 1 and October 31, basic wind speed of 115 mph (51 m/s) shall be permitted for structures sited near the Gulf Coast and Eastern Seaboard, where the ASCE/SEI 7-10 specified basic wind speed exceeds 115 mph (51 m/s) (3 second gust) provided additional bracing is prepared in advance and applied in time before the onset of an announced hurricane. The 115 mph (51 m/s) wind speed shall not be reduced by the factors in Section 6.2.1 for the construction period. The bracing shall be designed for the full, unmodified wind load determined using the mapped wind speed and procedures found in ASCE/SEI 7-10.

The basic wind speed used in this section is based on Risk Category II. If the authority having jurisdiction requires that the basic wind speed used in this section be based on Risk Category III or Risk Category IV, then the basic wind speed of 120 mph (54 m/s) shall be substituted for the basic wind speed of 115 mph (51 m/s) shown in this section.

C6.2.1.1.1 Construction Period in Hurricane-Prone Areas The dates selected to represent the hurricane season are not intended to include all times when hurricanes are possible. The dates are intended to include the period when the most severe hurricanes are probable.

If the construction site is in the path of a known oncoming hurricane, it is considered prudent to brace for the full, unmodified wind load determined using ASCE/SEI 7-10.

The changes to Section 6.2.1.1.1 were made to clarify the intent of the section. The previous edition was confusing and could have been interpreted to mean that the reduction could not be taken outside of the hurricane season. The change to the duration of the hurricane season in the context of loads on temporary structures is now more conservative.

6.2.1.2 Continuously Monitored Work Period For continuously monitored work periods, it shall be permissible to use wind speeds lower than those required in Section 6.2. For continuously monitored work periods, the basic wind speed shall be based on the speed, predicted by the National Weather Service or another reliable source acceptable to the authority having jurisdiction, for the day of construction. The basic wind speed shall not be less than the predicted wind speed adjusted to the 3-second gust speed, if necessary, multiplied by 1.26.

Continuously monitored work periods shall be those periods of continuous rigging, erection, or demolition that last for one work day or less. Continuously monitored work periods end at the end of the work day, at which time the structure shall be made inherently stable, or appropriately secured, to meet the requirements for the construction period as defined in Section 6.2.1.1.

C6.2.1.2 Continuously Monitored Work Period During erection, many structural components, including columns, girders, trusses, formwork, facade panels, etc., cannot be made to meet the requirements for wind resistance because they are being lifted or they have not been fully incorporated into braced and secured structures. Under such circumstances that last for one work day or less, it is permissible to use reduced wind speeds that are based on weather conditions predicted for the site. Temporary guys, struts, minimum number of fasteners, etc., should be employed as necessary for continuously monitored work periods. At no time should wind speeds used for continuously monitored work periods exceed those recommended by manufacturers of equipment used in the erection or demolition operation.

STANDARD | COMMENTARY

Weather forecasters sometimes publish predicted wind speeds based on sampling periods that are different from the 3-second gust that is the basis for wind loads in ASCE 7 [1]. The sampling period must be known, and the predicted wind speed must be adjusted to be consistent with provisions of ASCE 7 [1]. For the purposes of this standard, to obtain 3-second gust speeds multiply fastest-mile or one-minute average speeds by 1.20 and mean-hourly speeds by 1.53. The mapped wind speeds in ASCE 7[1] are intended for use with strength design methods. In order to convert the reported wind speeds to strength level wind speeds for use with a load factor of 1.0, the reported wind speeds must be multiplied by 1.26, which represents the square root of the wind load factor of 1.6 used in previous editions.

Certain rigging operations may by their very nature pose hazards, and may require more restrictive measures to conform to ordinances of the authority having jurisdiction, or to good practice. An example is to provide free-fall areas for the materials being handled, which may result in closure of streets or sidewalks, evacuation of buildings, etc.

6.2.2 Frameworks without Cladding Structures shall resist the effect of wind acting upon successive unenclosed components.

Treatment of staging, shoring, and falsework with a regular rectangular plan as trussed towers in accordance with ASCE/SEI 7-10 shall be permissible. Unless detailed analyses are performed to show that lower loads may be used, no allowance shall be given for shielding of successive rows or towers.

For unenclosed frames and structural elements, wind loads shall be calculated for each element. Unless detailed analyses are performed, load reductions due to shielding of elements in such structures with repetitive patterns of elements shall be as follows:

1. The loads on the first three rows of elements along the direction parallel to the wind shall not be reduced for shielding.
2. The loads on the fourth and subsequent rows shall be permitted to be reduced by 15%.
3. Wind load allowances shall be calculated for all exposed interior partitions, walls, temporary enclosures, signs, construction materials, and equipment on or supported by the structure. These loads shall be added to the loads on structural elements.

Calculations shall be performed for each primary axis of the structure. For each calculation, 50% of the wind load calculated for the perpendicular direction shall be assumed to act simultaneously.

C6.2.2 Frameworks without Cladding Even though the design wind speed during construction may be lower than that for the completed structure, the total wind load may actually be higher due to the cumulative effect of wind acting on many more surfaces and often with higher drag coefficients than in the fully enclosed structure. For common arrangements of elements in typical open frames and temporary structures, shielding effects are small. Considering the changing nature of the building silhouette and the arrangement of construction materials on the structure, it is prudent not to assume that loads will be reduced due to shielding, except in certain specific cases.

For open structures with regular patterns of elements, the direction of maximum force on the structure usually is not parallel to the principal axis of the structure. Shielding effects are minimized, and therefore loads are at their highest, when the direction of the wind is not parallel to the column lines. For this reason, the most severe loads on an open structure include components of load in both principal directions of the structure.

For guidance on shielding effects and loads on open structures, refer to Shapiro et al. (1999), Nix et al. (1975), Vickery et al. (1981), and the Metal Building Manufacturers Association (MBMA 1996).

6.2.3 Accelerated Wind Region Structures placed near building edges and corners shall resist the higher pressures and suctions that will exist in such regions. The design wind speed shall be factored upward from the basic wind speed by the square root of the suction coefficient for cladding as given in ASCE/SEI 7-10. The calculated wind

C6.2.3 Accelerated Wind Region Near building corners, at edges of completed building enclosures, and at other discontinuities in building geometry, the prevailing wind speeds are increased and wind directions are altered. Also, there may be substantial side and uplift loads on nearby adjacent structures (such as scaffolding) in these locations.

STANDARD

speed shall be used with appropriate drag factors to calculate loads on structures. At building corners, the resulting pressures shall be assumed to act on adjacent staging structures in horizontal directions parallel to and perpendicular to the enclosure surface. At top edges of enclosures, pressures shall be assumed to act upward as well as horizontally.

COMMENTARY

Special attention should be given to the loads on staging structures near the edges of enclosed and partially enclosed structures.

ASCE/SEI 7-10 and the Metal Building Manufacturers Association's *Low Rise Building Systems Manual* (MBMA 1996) provide guidance for pressures on edge regions of surfaces of enclosed structures, but there is little information on loads on such structures (e.g., staging constructed adjacent to and in the stream of air flowing around enclosed structures). Please be aware that ASCE/SEI 7-10 wind speeds are 3-second gust speeds and are intended for use with strength design methods using a load factor of 1.0. When using the MBMA's *Low Rise Building Systems Manual*, the ASCE/SEI 7-10 velocity pressure must be adjusted to the basis used in the MBMA's *Low Rise Building Systems Manual*. To adjust the ASCE/SEI 7-10 velocity pressure from a strength-based pressure to a working stress-based pressure, multiply the ASCE/SEI 7-10 velocity pressure by 0.6.

6.3 THERMAL LOADS

Provisions shall be made for thermal distortions of the structure and architectural components when structures are erected during the following conditions:

1. When the product of the following quantities exceeds 7,000 ft-°F (1,185 m-°C)
   a. The largest horizontal dimension between expansion joints of the erected structures; and
   b. The largest of the differences between the following temperatures for the months when the portion of the structure is erected and exposed temporarily to ambient temperatures:
      (1) The highest mean daily maximum temperature and the lowest mean daily minimum temperature; or
      (2) The expected average temperature of the structure when it is in its end use and the highest mean daily maximum temperature; or
      (3) The expected average temperature of the structure when it is in its end use and the lowest mean daily minimum temperature.
2. When portions of the structure which will be shielded when the structure is completed are subjected to direct solar radiation during hot weather; or
3. Whenever temperature changes create distortions that could damage structural or architectural components.

C6.3 THERMAL LOADS

Thermal distortions can be significant when frames of structures under construction are exposed to daily or seasonal ambient temperature variations, and when frames are erected at temperatures that differ significantly from the temperature of the structure in its end-use condition.

The provisions of this section limit theoretical structural distortion between expansion joints to approximately 0.5 in. (13 mm). Thermal lag of structural elements is considered by specifying that calculations be based on highest mean daily maximum and lowest mean daily minimum temperatures for the months when the structure is exposed to ambient temperatures.

The National Climatic Center's *Climatological Summary of the U.S.* (NCC1) contains mean daily maximum and minimum temperatures that are suitable for use in thermal load evaluations. Additional temperature data are in the Federal Construction Council's *Technical Report No. 65* (Building Research Advisory Board 1974) and the National Climatic Center's *Monthly Normals of Temperature, Precipitation, Heating and Cooling Degree Days* (NCC2).

When the attachment of structural elements to foundations and adjacent structures is flexible, thermal distortions can result in movement that causes forces in those components and attachments without serious consequences. However, if the attachments are rigid, extremely large forces may develop because of the restraint of movement. Damage occurs when stressed elements are incapable of supporting the resulting forces.

Although damage is possible in almost any building, those that are most susceptible have relatively unrestrained frames supporting rigid elements, such as precast panels or masonry infilling walls, which are not a part of the primary structural system (Martin 1971). Long buildings, in which the cumulative dimensional changes can be large, and buildings erected during the extremes of the construction season when ambient temperatures can be very different from end-use temperature, are particularly susceptible. Also, structures with braced bays or shear walls in line but spaced far apart can generate substantial forces as the intermediate framing attempts to move with temperature changes.

Multistory buildings usually show the most damage in the lowest stories (ACI 1985), where the foundation provides the greatest restraint to free movement.

Some components can develop substantial flexural distortions and/or forces due to solar radiations on a large surface during construction (Martin 1971; Chrest et al. 1989; ACI 1992; PCI 1992; Ho and Liu 1989); this can be detrimental for a component that is designed to be shielded in the finished structure.

Thermal distortions are often impossible to restrain because the forces that are generated exceed the capacities of practical restraining elements. Therefore, it is advisable to accommodate distortions by sequencing the erection so as to avoid making rigid connections between portions of the structure that may undergo differential movement until the temperature of the frame can be stabilized, or by installing structural and architectural details that will tolerate movement.

6.4 SNOW LOADS

When snowfall is expected during the construction period, as defined in Section 6.2.1.1, except as modified herein, snow loads shall be determined for surfaces on which snow could accumulate in accordance with ASCE/SEI 7-10. If construction will not occur during winter months when snow is to be expected, snow loads need not be considered, provided that the design is reviewed and modified, as appropriate, to account for snow loads if the construction period shifts to include winter months.

Design for snow loads that are lower than those prescribed by this section shall be permissible, provided adequate procedures and means are employed to remove snow before it accumulates to levels that exceed the loads used for design.

C6.4 SNOW LOADS

6.4.1 Ground Snow Loads The ground snow loads, p_g, given in ASCE/SEI 7-10 shall be modified by the following factors:

Construction Period	Factor
Five years or less	0.8
More than five years	1.0

C6.4.1 Ground Snow Loads In recognition of the relatively short duration of most construction projects, the ground snow load is reduced for durations of five years or less to reflect the low probability that the 50-year mean recurrence interval value, which forms the basis for ASCE/SEI 7-10 loads, will occur during the construction period. However, it should also be realized that the loads in excess of statistically determined design loads may occur.

6.4.2 Thermal, Exposure, and Slope Factors The thermal factor, C_t, and the exposure factor, C_e, shall be for the conditions that will exist during construction. If a range of conditions will exist during construction, a series of load calculations shall be made to cover the range of thermal

C6.4.2 Thermal, Exposure, and Slope Factors The values of the thermal factor, C_t, are determined for the conditions that will exist during construction. These conditions may be quite different from those that will exist once the building is occupied. Because the value of C_t depends

STANDARD

and exposure factors to be expected. The slope factor, C_s, shall be determined based on the construction-phase value of C_t.

6.4.3 Drainage Where drainage system components may become blocked during construction (e.g., by freezing), the extra loads created by such blockages shall be included.

6.4.4 Loads in Excess of the Design Value Surfaces on which snow and ice accumulate shall be monitored and any loads in excess of construction-phase design loads shall be removed before construction proceeds.

6.5 EARTHQUAKE

If required by Section 6.5.1 and not exempted by Section 6.5.3, earthquake loads shall be calculated in accordance with procedures in ASCE/SEI 7-10 as modified by Section 6.5.2. All structures shall be treated as Risk Category II, per Table 1.5-1 of ASCE/SEI 7-10, regardless of the group classification of the completed structure.

6.5.1 Applicability Earthquake loads need not be considered unless required by the authority having jurisdiction and the mapped Risk-Targeted MCE_R, 5% damped, spectral response acceleration parameter at a period of 1 s, S_1, defined in Section 11.4.1 of ASCE/SEI 7-10 equals or exceeds 0.40.

Construction of detached one- and two-family lightly framed dwellings not exceeding two stories in height is exempt from these earthquake requirements.

This section applies to all construction except those specifically covered in Section 6.5.3.

6.5.2 Use of ASCE/SEI 7-10 For use of the earthquake load provisions of ASCE/SEI 7-10, the following modifications should be made:

1. The mapped values for S_S and S_1 may be multiplied by a factor less than 1 to represent the reduced exposure period, but the factor shall not be less than 0.20.
2. The restrictions on types of structural systems in seismic performance categories D and E do not apply, as long as the height of the temporary bracing system designed in accordance with this section is limited in height to 60 ft (18.3 m) or five stories,

COMMENTARY

on whether heat is provided in a building, and most buildings under construction are unheated, snow loads may be higher during construction than when the building is completed and occupied.

In most circumstances the exposure factor, C_e, for a roof during construction will be the same as the exposure factor for that roof during the life of the building. When the thermal factor changes, the slope factor, C_s, may also change.

C6.4.3 Drainage Drainage system components, which often rely on building heat to function properly, may become blocked with ice during construction in cold weather if the building is unheated. When this occurs, excess loads may accumulate on roofs. Ponding instability may result. It may be appropriate to install temporary heaters in drains to avoid such problems during construction.

C6.4.4 Loads in Excess of the Design Value If loads in excess of construction-phase design values are encountered during construction, work within the building should be halted until the overload is eliminated. Snow removal procedures must be planned to avoid overloading the structure with piles of snow or by the use of equipment too heavy for the structure.

C6.5 EARTHQUAKE

The earthquake provisions of ASCE/SEI 7-10 are modeled on the 2009 *NEHRP Recommended Provisions for Seismic Regulations for New Buildings and Other Structures* prepared by the Building Seismic Safety Council (2009).

C6.5.1 Applicability It is not reasonable to require seismic resistance for temporary works where large earthquakes are infrequent or not considered probable.

C6.5.2 Use of ASCE/SEI 7-10 The ground motion response acceleration with a 2% chance of exceedance in one year (a mean recurrence interval of about 50 years) (S_s*) ranges from approximately 5% to 20% of the value mapped as having a 2% chance of being exceeded in a 50-year period (a mean recurrence interval of about 2,475 years). The percentage is in the upper portion of the range in the more seismically active areas. A value of 20% is selected as representative of this ratio. The intent of this section is for the soil amplification factors, F_a and F_v, to be based on the full value of S_S and S_1. With the introduction

STANDARD

whichever is less, above the completed bracing of the permanent structure.

3. The R factor used for temporary bracing systems shall not exceed 2.5 unless the system is detailed in accordance with the provisions of ASCE/SEI 7-10. Where R = 2.5 is used, only the requirements dealing with the strength of the seismic-resisting structural system need be satisfied.

6.5.3 Other Standards for Earthquake-Resistant Design Partially complete structures of types that are excluded by the earthquake load provisions of ASCE/SEI 7-10 and for which specifically applicable standards for earthquake-resistant design exist, such as vehicular bridges, shall be designed and evaluated according to the specifically applicable standard. Earthquake loads for temporary structures associated with such construction shall be determined in accordance with Section 6.5.2 unless the specifically applicable standard includes provisions for temporary construction.

6.6 RAIN

Except as modified herein, rain loads shall be calculated in accordance with procedures in ASCE/SEI 7-10.

For temporary conditions that exist for one month or less, rain loads need not be considered for construction during months with historical rainfall averages of less than 1 in. (25 mm) per month.

COMMENTARY

of ASCE/SEI 7-10, the basis of the ground motions changed from a 2% chance of being exceeded in a 50-year period to a target risk of collapse of 1% in 50 years. Although the basis of the ground motions used in ASCE/SEI 7-10 has changed, the ratio of these values based on 1 year and 50 years does not change.

The drift limitations and the nonstructural provisions are not required for temporary structures and for structures during their construction phases.

Specific provisions for design of masonry structures to resist seismic loads are contained in TMS 402-11/ACI 530-11/ASCE 5-11 (MSJC 2011).

Designers of temporary bracing systems taller than 60 ft (18.3 m) or five stories should use rational approaches that are acceptable to the authority having jurisdiction.

C6.5.3 Other Standards for Earthquake-Resistant Design ASCE/SEI 7-10 excludes certain types of structures for various reasons: unique characteristics of response to ground shaking, exceptionally high risk associated with poor performance, and the existence of other standards for design. The provisions of Section 6.5.2 are intended to make ASCE/SEI 7-10 usable for most temporary structures; however, it is beyond the scope of this document to repeat complete sets of seismic design provisions for structures already covered by existing standards. For vehicular bridges, the user is referred to the provisions of AASHTO (2002) and CALTRANS (2004a). Also, CALTRANS *Memo to Designers* 20-2 (1989), 20-12 (2003), and 15-14 (2004b) recommend acceleration levels to use for temporary situations involving bridges carrying traffic or positioned over traffic, that are higher than the values specified in paragraph 1 of Section 6.5.2 for near-fault locations.

C6.6 RAIN

In some regions of the country, seasonal rainfall is very low. For construction in these regions during low-rainfall seasons, it is not essential to consider rain loads. Nevertheless, water should be removed when it accumulates in or on structures to a sufficient depth to exceed 25% of the live, rain, or snow loads on any supported structural element as specified in this standard.

Many structures drain better while under construction than when they are finished. However, there are circumstances when drainage potential is reduced. An example is an unfinished parking deck which relies on a sloped topping slab for drainage. In this case, permanent drains might be above the level of finished construction and the surface of the slab might be essentially level. Also, drainage systems can become blocked with ice during freezing conditions (see Section 6.4.3) and construction debris.

Care must be taken to keep drains clear and to provide for unobstructed paths for rain water to flow from structures. Water that accumulates in unfinished structures should be removed.

STANDARD

6.7 ICE

Except as modified herein, ice loads shall be calculated in accordance with procedures in ASCE/SEI 7-10.

For construction during seasons when structures are not susceptible to accumulation of ice, ice loads need not be considered.

Structures which will be enclosed when construction is complete and which are designed for live loads of 20 psf (0.96 kN/m^2) or more need not be considered as ice-sensitive structures while open during construction.

COMMENTARY

C6.7 ICE

Enclosed structures which are designed for significant live loads on floor areas need not be considered for ice loads solely because they have an open configuration during construction. However, should ice accumulate on these structures, it should be removed or the construction and live loads applied to the structure should be reduced by an amount corresponding to the weight of the accumulated ice.

REFERENCES

American Association of State Highway and Transportation Officials (AASHTO). (2002). *AASHTO standard specifications for highway bridges*, 17th Ed., Washington, DC.

American Concrete Institute (ACI). (1985). "State-of-the-art report on temperature-induced deflections of reinforced concrete members." *ACI 435.7R-85*, Farmington Hills, MI.

ACI. (1992). "Analysis and design of reinforced concrete guideway structures." *ACI 358.1R-92*, Farmington Hills, MI.

ASCE/SEI. (2010). "Minimum design loads for buildings and other structures." *ASCE/SEI 7-10, Including Supplement No. 1*, Reston, VA.

Boggs, D. W. and Peterka, J. A. (1992). "Wind speeds for design of temporary structures." *Proc., ASCE 10th Structures Congress*, San Antonio, TX, April 13–15, 1992, ASCE, Reston, VA, 800–803.

Building Research Advisory Board, Federal Construction Council. (1974). "Expansion joints in buildings." *Tech. Rep. No. 65*, National Academy of Sciences, Washington, DC.

Building Seismic Safety Council. (2009). "NEHRP recommended provisions for seismic regulations for new buildings and other structures." *FEMA P-750 2009*, Federal Emergency Management Agency, Washington, DC.

California Department of Transportation (CALTRANS). (1989). "Seismic requirements for staged construction." *Memo to Designers, November 1989, 20-2*, Sacramento, CA.

CALTRANS. (2003). "Site seismicity for existing and temporary bridges carrying public vehicular traffic." *Memo to Designers, February 2003, 20-12*, Sacramento, CA.

CALTRANS. (2004a). *Bridge Design Specifications, September 2004*, Sacramento, CA.

CALTRANS. (2004b). "Loads for temporary highway structures." *Memo to Designers, February 2004, 15-14*, Sacramento, CA.

Chrest, A. P., Smith, M. S., and Bhuyan, S. (1989). *Parking structures*, Van Nostrand Reinhold, New York, 141.

Ho, D., and Liu, C.-H. (1989). "Extreme thermal loadings in highway bridges." *J. Struct. Eng.*, 115(7), 1681.

Martin, I. (1971). "Effects of environmental conditions on thermal variations and shrinkage of concrete structures in the United States." *Designing for effects of creep, shrinkage, and temperature in concrete structures, SP-27*, American Concrete Institute, 289–299.

Mason Contractors Association of America (MCAA). (2001). *Standard practice for bracing masonry walls under construction*, Lombard, IL. (Also available from the National Concrete Masonry Association, Herndon, VA as *TR-171-2001*.)

Masonry Standards Joint Committee (MSJC). (2011). "Building code requirements for masonry structures." *TMS 402-11/ACI 530-11/ ASCE 5-11*. Joint publication of The Masonry Society, American Concrete Institute, and Structural Engineering Institute of ASCE, ASCE, Reston, VA.

STANDARD

COMMENTARY

Metal Building Manufacturers Association (MBMA). (1996). *Low rise building systems manual*, Cleveland, OH, 223–226.
National Climatic Center (NCC1). *Climatological summary of the U.S.* Asheville, NC.
NCC2. *Monthly normals of temperature, precipitation, heating and cooling degree days*, Asheville, NC.
Nix, H. D., Bridges, C. P., and Powers, M. G. (1975). "*Wind loading on falsework, Part I.*" California Department of Transportation, Sacramento, CA.
Prestressed Concrete Institute, Committee on Parking Structures (PCI). (1992). *Precast prestressed concrete parking structures: Recommended practice for design and construction*, Chicago.
Ratay, R. T. (1987). "To mitigate wind damage during construction: Codification?" *Proc., NSF/Wind Engineering Research Council Symp. on High Winds and Building Codes*, Kansas City, MO, November 2–4, 1987.
Rosowsky, D. V. (1995). "Estimation of design loads for reduced reference periods." *Struct. Safety*, 17, 17–32.
Shapiro, H. I., Shapiro, J. P., and Shapiro, L. K. (1999). *Cranes and derricks*, 3rd Ed., McGraw-Hill, New York, 144–178.
Vickery, B. J., Georgiou, P. N., and Church, R. (1981). "Wind loading on open framed structures." *Proc., 3rd Canadian Workshop on Wind Engineering*, Vancouver and Toronto, Canada, 1981.

INDEX

Page numbers followed by *e*, *f*, and *t* indicate equations, figures, and tables, respectively.